LABORATORY EXERCISES IN DEVELOPMENTAL BIOLOGY

LABORATORY EXERCISES IN DEVELOPMENTAL BIOLOGY

Yolanda P. Cruz

Department of Biology
Oberlin College
Oberlin, Ohio

ACADEMIC PRESS
A Division of Harcourt Brace & Company
San Diego   New York   Boston   London   Sydney   Tokyo   Toronto

Academic Press, Inc.
1250 Sixth Avenue, San Diego, California 92101-4311

United Kingdom Edition published by
Academic Press Limited
24–28 Oval Road, London NW1 7DX

Laboratory Exercises in Developmental Biology
by Yolanda P. Cruz
International Standard Book Number: 0-12-198390-0

PRINTED IN THE UNITED STATES OF AMERICA
93 94 95 96 97 98   MM   9 8 7 6 5 4 3 2 1

CONTENTS

Preface   ix

Background   1

EXERCISE **1**
Fertilization in the Sea Urchin   11

EXERCISE **2**
**Early Development in the Sea Urchin
and Starfish**   23

EXERCISE **3**
**Early Development in the Frog:
To Neurulation**   35

EXERCISE **4**
**Early Development in the Frog:
To 10 Millimeters**   49

EXERCISE **5**
**Preventing Axis Formation in
Amphibians with Ultraviolet Irradiation**   65

EXERCISE 6
Early Chick Development: To 48 Hours   81

EXERCISE 7
Early Chick Development: To 96 Hours   97

EXERCISE 8
Chick Development in a Windowed Egg   111

EXERCISE 9
**Ethyl Alcohol, Caffeine, and
Chick Development**   123

EXERCISE 10
**Totipotency of the Unincubated
Chick Blastoderm**   137

EXERCISE 11
Cardia Bifida in the Chick   147

EXERCISE 12
Tadpoles and Thyroxine   161

EXERCISE 13
**Blastomere Totipotency in the
Two-Cell Mouse Embryo**   175

EXERCISE 14
Alizarin Staining   189

CONTENTS

EXERCISE **15**
Drosophila Polytene Chromosomes 199

EXERCISE **16**
**Induction of Chromosome Puffing
with Heat Shock** 213

EXERCISE **17**
Drosophila Imaginal Discs and Ecdysone 223

Index 235

PREFACE

Developmental biology is a rapidly expanding field of study. From its origins in classical embryology and experimental morphology, developmental biology has come to include, in recent years, many facets of genetics, cell and molecular biology, biochemistry, and comparative anatomy. Keeping up with new knowledge in developmental biology has become an almost daunting task, leaving many instructors with no choice but to concentrate on incomplete but familiar subsets of topics. Because of this, teaching a course in developmental biology has become a real challenge.

An additional challenge faces those of us who teach laboratory courses in developmental biology at the college level. There is the constant dilemma of deciding which laboratory exercises would be appropriate and manageable within the constraints of limited time, space, and material resources. Additionally, there is legitimate concern about the learning benefits of those laboratory exercises we eventually select. Most college instructors in developmental biology agree that a firm grounding in embryology is paramount, but that familiarity with novel experimental approaches to problems in developmental biology is no less worthwhile. Doing justice to the discipline while meeting the needs of our students then becomes a potentially arduous exercise in striking just the right balance.

This manual is intended as an accompaniment to a lecture course in animal development. Its focus is on fostering familiarity with the different animal experimental systems that have provided us with opportunities to understand just what is going on at the organismal, cellular, and subcellular levels. Its approach is a combination of learning techniques, performing experiments, and providing opportunities to formulate questions about interesting problems in developmental biology. Inherent in its content and organization is the recognition that neither observation nor experimentation alone suffices as individually adequate strategies in comprehending recent advances in animal development. The manual deliberately takes a broad, introductory view of developmental systems and thus, of necessity, sacrifices some depth in the treatment of any one of these experimental systems. The breadth in coverage that is gained, however, more than convincingly compensates for the exclusion of laboratory exercises whose main focus would be narrower, or would be more appropriate for a class in cell or molecular biology. Indeed, the laboratory exercises presented here are best described as laboratory exercises in animal development.

This volume is the result of 6 years of teaching developmental biology to college juniors and seniors at Oberlin College. With rare exceptions, these students have academic interests in biology, neuroscience, and biochemistry. Without exception, these students have had introductory courses in organismal biology and in cell and molecular biology. This manual is designed for a class of 20 meeting twice weekly. Many of the exercises can be completed during a 3-hour lab; some span the course of days; others, weeks. The sequence in which the exercises appear here is dictated by the lecture topics. Each exercise, however, can stand on its own; for this reason, a section called "Notes to the Preparator" is included in every exercise.

Many students and colleagues gave advice and assistance during the formative years of these lab exercises. I am forever in their debt: Steve Black, Dennis Chang, Kirsten Hagstrom, Edina Harsay, Anne Hastings, Roger Pedersen, Johann Soults, Jim Sutherland, and Cheryl Wolfe. I am grateful to my students in Biology 302 for their comments and suggestions; to my colleague, Robin Treichel, and my editor, Phyllis Moses, for encouragement. I thank Elsa Pearson and David Love for their forbearance. Finally, I acknowledge the helpful suggestions offered by several anonymous reviewers.

Yolanda P. Cruz

Background

INTRODUCTION

The scientific process is frequently perceived as a smooth execution of flawless protocols from which great insights and breathtaking revelations result. Indeed, a reading of scientific papers reinforces this view. However, there is perhaps no greater myth than this. The scientific process is fraught with pitfalls, unforeseen problems, and unsuccessful attempts — the majority of which go unreported and thus, unappreciated. But this is precisely what makes it a process: It is an unending search for answers. Beyond this, precise descriptions of "the scientific process" are difficult. Its most predictable attribute is perhaps its unpredictability.

So, how does one become proficient in the scientific process? What makes a scientific investigator successful? Familiarity and experience are good starting points. Patience is another. A willingness to evaluate mistakes and learn from them is certainly mandatory, as is the determination to try yet again. Finally, a scrupulous adherence to honesty, thoroughness, and candor helps, too. It's a long trek to there. Let's get started.

The laboratory exercises in this manual are designed to familiarize you with common techniques and organisms used in the study of animal development. They are intended to help in developing a systematic approach to assessing problems and devising means by which such problems may be solved. Therefore, in certain exercises, you will be observing and comparing; in others, experimenting. In all exercises, however, you will be constantly recording and evaluating what you have accomplished, interpreting your results, analyzing their implications, and formulating further questions for future work.

The usefulness of this manual depends on its user. A careful reading of the exercise to be performed *before* coming to class is a most effective way of minimizing mistakes and avoiding frustration. Making a written outline of the protocol and a list of required equipment and materials on the note pages provided will improve performance of that exercise. Supplementary readings are provided with every exercise; questions are posed to evoke answers drawn from material presented in both the laboratory and the lecture. The student is asked to write reports to be handed in for grading. In this connection, keeping laboratory notes will be helpful for careful recording of observations and meticulous documentation of work done in the laboratory. The notes will serve as a chronicle not only of your activities but also of your impressions, errors, inspirations, and flashes of insight. The note pages provided with each exercise could be profitably used for this purpose.

The lab reports, handed in soon after the conclusion of each exercise, should indicate the objectives of the exercise, report the information in the form of drawings, graphs, and tables, and present a summary of your conclusions, including explanations for various novel or unanticipated observations, possible causes of errors, and flaws in experimental design. Conclude each report by asking a question provoked by the lab exercise just completed.

One last thing: For some of the lab exercises, you will be working individually; for others, in pairs or small groups. Take full advantage of the interactions you will have with your classmates by discussing your results with them, sharing responsibilities, and listening to suggestions. Creative input and cooperation clearly enhance mastery of the scientific process.

USING THE COMPOUND MICROSCOPE

The lighting in a microscope must be set up properly to maximize resolution. Given here is a list of steps to follow. If you need to refresh your memory about the names and functions of the various parts, consult the operator's manual, which instructors should make available in the laboratory.

For this class, you need not be concerned with the physics of microscopy, but you will need to know what a microscope does and how to take full advantage of its capabilities by becoming familiar and comfortable with this scientific instrument.

The microscope is equipped with an illumination source at the base. Located beneath the *microscope stage* is a *field diaphragm* and a *substage condenser diaphragm.* The condenser does just that: it condenses or collects light, concentrating the light as it goes through the specimen. The wider the cone of light that passes through the specimen, the greater the resolution (that is, the ability to distinguish objects or images close together). Thus, if the *field diaphragm* is closed down too far, resolution will be poor. To improve contrast (or light–dark diversity of adjacent parts), adjust the *substage diaphragm.*

1. The microscope should have its *low-power objective lens* in place and the *stage* cranked down as far as it will go. If not, then make the necessary adjustments.

2. Make sure both *diaphragms* are open.

3. With the light intensity knob at zero, turn on the *illuminator.* (At brighter settings, the bulbs get "shocked," and so do your eyes.) *Then,* gradually turn up the light — but be conservative.

4. Adjust both diaphragms by bringing them into focus with a practice microscope slide.

5. Adjust the height of the *condenser* to optimize illumination.

6. Adjust the distance between the *eyepieces* or *oculars,* as well as the distance of each *ocular* from each of your eyes. (This is the difference between comfort and migraine.)

7. If your microscope has a *zoom knob,* note how this feature allows you to increase magnification slightly without changing *objective lenses.*

8. Now, increase magnification by changing *objective lenses.* (Grab the *nosepiece,* not the *objective.*) When using this procedure to increase magnification, close down the *field diaphragm* to just the limit of the visual field of the objective in use.

9. Adjust or reset your *substage diaphragm.*

Reminders on Microscope Use

1. Treat your microscope with respect and care. Pick it up by grabbing the *arm* with your dominant hand and supporting the *base* with your other hand. This will reduce the possibility of banging or losing parts that could fall off or out.

2. Do not force levers or knobs. If something feels funny or seems to be reluctant, STOP! Get help.

3. Do not remove parts of the microscope, or open ocular backs or objective backs. This allows dust and moisture in.

4. Use *lens tissue* to clean *external* glass surfaces. (No other tissue papers are suitable.) Do not use solvents for cleaning. Remember to wipe dry any microscope part with oil, water, or any other liquid on it.

5. Avoid wetting or smearing the stage, because this will corrode it.

6. Use the dust covers. Return the microscope and illuminator to the desk cabinet, unless instructed to do otherwise.

WRITING SCIENTIFIC PAPERS

Scientific papers have unique characteristics. First, they are written mainly for other scientists; this allows the author to assume a baseline level of knowledge in the reader. Second, the language of scientific papers is often dry and dense. Reading, therefore, goes slowly. Writing proceeds even more slowly; the author is forced to write, edit, and rewrite, over and over again. Frequently, an author will request colleagues to review a paper prior to submitting it for publication. Journal editors will typically demand that submitted manuscripts be as concise as possible; in this connection, many journals charge per-page costs beyond a maximum number of pages that are printed free of charge. The goal in writing a scientific paper is to report experimental or observational results; scientific papers are thus meant to be read. It is the author's responsibility to make the paper clear, concise, and readable.

The following paragraphs list the major parts of a manuscript and describe what the contents for each should be. Choose a paper from a recent scientific journal and examine how each of the following sections is written. Every journal has its peculiarities. Regularly, an issue appears in which "Instructions for Contributors" will be printed. These instructions are meant to be followed to the letter.

The ABSTRACT is a miniature version of the paper itself. Thus, it contains a sentence or two that introduces the topic, describes the hypothesis, details the materials and procedures used, reports the results, summarizes the discussion, and ends with a conclusion. The reader needs only to read the abstract to understand what the paper is all about.

The INTRODUCTION puts the topic in proper perspective and gives a brief history or background of the problem. It includes literature reviews, states the hypothesis being tested, and clearly delineates the intent of the paper.

The MATERIALS AND METHODS section is best written with subheadings that reflect groupings of related protocols. For instance, in the sea urchin fertilization lab, appropriate subheadings could be "Procurement of Gametes" (or something along those lines) and "Detection and Timing of Fertilization Envelope Formation." Remember that the reader should be able to repeat exactly what the author did by simply reading this section of the paper.

The RESULTS section reports the findings. Data may be presented in the text, or in figures, tables, and graphs. No given data set should ever be presented in more than one form. It is a good idea to make a list of the new findings being reported. Developing a logically structured outline from this list then becomes easy. No interpretation of data should be included in this section.

The DISCUSSION section should be devoted to examining the implications of the data, and not to re-reporting the data. This section is appropriate for putting the present results into the greater perspective of what is currently known in the same research area. Comparisons, analyses, extrapolations, contradictions, interpretations—all these belong in this section.

The SUMMARY or CONCLUSIONS section is optional. A summary reports only what was done (unlike an abstract) and the conclusions obtained. A conclusion section essentially does the same thing, but focuses more tightly on what conclusions were provoked by the data.

The rules of good writing remain in force in the writing of a scientific paper. For instance, every paragraph should be well composed, complete with a topic sentence and an ending remark. Editing and re-editing pays off with a paper that will read smoothly. Working from an outline is always a good idea because it allows you to organize the material you wish to present and be thorough in your treatment of it. Well-written papers owe their readability to the author's fastidious adherence to these guidelines.

EQUIPMENT ASSIGNMENT

How many	What	Identifying number or description
	Compound microscope	
	Accompanying transformer	
	Optical filters	
	Eyeguards	
	Dissecting microscope	
	Illuminator	
	Slide box and slides	

I assume full responsibility for the proper care and use of the above items while they are assigned to me. I further assume responsibility for reporting any defect or malfunction immediately.

____

Student's Signature

Above Equipment Returned: _______________________________

Date

Remarks:

____

Releaser's Name

NOTES

NOTES

Fertilization in the Sea Urchin

The sea urchin, an echinoderm, is a classic embryological specimen. A male urchin releases several billion sperm; the female releases about one billion eggs *each* time gamete-shedding is induced. Induction is simple; a small volume of potassium chloride (KCl) injected into the body cavity of the sea urchin causes it to release gametes within minutes. The gametes are translucent and small enough to be observed under the microscope. The cytoplasm is clear, and cleavage and gastrulation are easily observed.

Fertilization is *external* in the sea urchin. As in other organisms, this event represents the union of the male and female gamete into a *zygote*. Immediately before this union, the sperm undergoes an *acrosome reaction* during which its acrosome (fused lysosomes) releases lytic enzymes that help to penetrate the egg's *jelly coat*. Almost at the same time, the sperm produces a long filament from its cell membrane, or *plasmalemma,* which fuses with the plasmalemma of the egg. This fusion quickly produces a transient electrical depolarization across the egg plasmalemma, which loses its ability to fuse with other sperm. This event constitutes the *initial block to polyspermy.*

The *late* (or permanent) *block to polyspermy* is the *fertilization envelope.* This structure consists of the congealed contents of the egg's *cortical granules,* which fuse with the egg's plasmalemma during sperm entry and thus release their contents into the immediate vicinity of the egg. The fertilization envelope lifts (or peels) away from the surface of the egg plasmalemma, an event that takes only seconds and therefore requires close observation.

In this exercise, you will (1) determine how sperm concentration affects fertilization efficiency; (2) view the formation of the fertilization envelope first with the addition of sperm only to an egg suspension and then with a calcium ionophore only; and (3) examine the effects of egg-jelly extract on the ability of sperm to fertilize oocytes. You will be working in groups of three or four. Each group is responsible for dividing the work required by this exercise and for submitting a jointly written report in the form of a scientific paper. Results will be discussed in class.

During the first experiment, you will determine the minimum sperm concentration that effects good fertilization and use this value to view and describe (in detail) the formation of the fertilization envelope (second experiment). In the second experiment, you will also examine the formation of the fertilization envelope *when only A23187 (a calcium ionophore), but no sperm, is used.* Be careful that all containers used for this trial are uncontaminated by sperm. Finally, for the third experiment, you will dissolve sea urchin egg jelly by decreasing the pH of the seawater in which the eggs are suspended. You will then expose sperm to the resulting egg-jelly extract. Fresh sperm exposed only to egg-jelly extract undergo the acrosome reaction. Compare the fertilization efficiency of extract-exposed sperm with that of sperm exposed only to acidified seawater.

PROCEDURE

Read the following procedure first. Assemble the materials needed before beginning the actual experiments.

1. Obtain a test tube rack with 4 test tubes, 1 or 2 depression slides, and 5 Pasteur pipettes with rubber bulbs. Assign a Pasteur pipette to each test tube; mark the test tubes "Eggs," "1," "2," and "4," respectively. Set up *both* the dissecting and compound microscopes. You should have a stopwatch, timer, *or* a wristwatch with a second hand.

2. Place 5–10 ml of egg suspension in the "Eggs" test tube. The labels on the other tubes refer to the concentration of the sperm suspension (drops of dry sperm per 10 ml seawater) to be used in evaluating the effects of sperm concentration on the rate of fertilization. These suspensions will be dispensed by the instructor and should be used within 10 min.

3. Place a drop of egg suspension on your depression slide. View under the dissecting scope. Cover the well of the slide with a coverslip when examining under the compound scope. Make a sketch of unfertilized oocytes.

4. Clean the slide by washing it with deionized water (no soap), then wiping it dry with Kimwipes sheets. Prepare test tube 1 by adding 1 drop of "dry" (undiluted) sperm to it.

5. Be ready to activate the timer. Combine 1 drop of egg suspension and 1 drop of the recently prepared sperm suspension in the well of your slide *while looking through the dissecting microscope.* Start timing as soon as you have combined the two drops of gametes. *Keep your eyes on the microscope field.*

6. As soon as the first fertilization envelope lifts off the surface of a fertilized oocyte, record the seconds or minutes elapsed since the drops of gametes were combined. Immediately reset the stopwatch or timer.

7. With your eyes still on the microscope field, and with a partner keeping time, count the number of fertilization envelopes that will form in the next 10 sec. Record this number as a fraction: the number of fertilized eggs divided by the total number of eggs in the microscope field.

8. Repeat Steps 4 through 7 twice, using the other concentrations of sperm.

9. Tabulate the results on a grid of your design in your notebook or in the note pages provided and on the blackboard with the rest of the class.

10. Make a slide mount of sperm (use the test tube labeled "4," containing the suspension with 4 drops of sperm) with a regular glass slide and coverslip. A small drop will do. Examine under the compound microscope. Make a sketch of one or two sperm. Remember to measure the sperm head and tail lengths.

11. Discard used coverslips and Pasteur pipettes in the glass-disposal bin. Wash the depression slides and allow them to dry. Do not use soap on the slides. Return rubber bulbs to the designated containers. Before putting away the microscopes, wipe dry the stages and lenses of both oculars (or eyepieces) and objective lenses with lens paper.

12. Now, you are on your own. Using this basic procedure, perform the experiments described in the introduction section to this lab exercise. We will use A23187 as follows: Dissolve 1 mg of A23187 in 1 ml dimethyl sulfoxide (DMSO), then use 1 part of this solution per 100 parts seawater.

13. In your report, present your results and discuss their implications. Consider making graphs and tables to present data not presented in the text. Address, among others, the following questions: (1) What was the final concentration of A23187 used? (2) What could be the role of A23187 in the formation of fertilization envelopes in the absence of sperm? (3) What does this reveal about the spermless activation of oocyte development (artificial parthenogenesis)?

14. To obtain egg-jelly extract, use a microcentrifuge tube initially filled with 1 ml of egg suspension to obtain an egg pellet 5–10 mm deep. Remove as much of the supernatant seawater as possible by careful aspiration with a Pasteur pipette.

15. Resuspend the eggs in about 3 or 4 volumes of seawater previously acidified to pH 4.5 with acetic acid. Manually swirl the microcentrifuge tube for 5 min, then microcentrifuge the eggs into a pellet. Transfer the supernatant, which should contain the egg-jelly extract, into a new tube. Label this microcentrifuge tube accordingly.

16. Obtain another microcentrifuge tube containing acidified seawater. Using dilute sodium hydroxide (NaOH) in seawater and a pH meter, titrate the pH in this tube and in the egg-jelly extract tube to pH 7.1. Cool down both tubes to 15°C before using.

17. Expose sperm to egg-jelly extract by putting one drop of "dry sperm" into 25 ml of cold seawater and stir by swirling the container several times. Immediately add 1 drop of this sperm suspension to 1 drop of egg suspension on a depression slide.

18. Count the number of fertilization envelopes that form. Repeat this step using acidified seawater only. Does pretreatment with egg-jelly extract affect the ability of sperm to fertilize oocytes? Explain your answer in your report.

For Your Reference

Approximate time sequence for the development of fertilized sea urchin eggs.

Formation of fertilization envelope	2–5 min
First cleavage	50–70 min (approx. 1 hr)
Second cleavage	78–107 min (approx. 1 hr 20 min to 1 hr 45 min)
Third cleavage	103–145 min (approx. 1 hr 45 min to 2 hr 25 min)
Blastula	6 hr
Hatching	7–10 hr
Gastrula	12–20 hr
Pluteus	24–48 hr

QUESTIONS

1. Propose an adaptive significance for the large number of gametes produced and released by sea urchins.

2. Briefly describe the changes in the egg upon the addition of sperm to the egg suspension.

3. Did the rate of fertilization envelope formation increase with sperm concentration? Why or why not?

4. Did any fertilization envelopes form when oocytes were placed in seawater containing A23187? Explain your results.

5. How well did sperm pretreated with egg-jelly extract compare with untreated sperm in terms of inducing the formation of fertilization envelopes? Explain.

References

Dan, J. C. (1952). Studies on the acrosome. I. Reaction to egg and water and other stimuli. *Biol. Bull.* **103,** 54–66.

Gilbert, S. F. (1991). *Developmental Biology,* 3rd ed, pp. 41–46, 51–58. Sunderland, Massachusetts: Sinauer Associates.

Giudice, G. (1973). *Developmental Biology of the Sea Urchin Embryo,* pp. 63–86. New York: Academic Press.

Merriam, R. W. (1988). *Experiments in Animal Development,* pp. 10–15. Sunderland, Massachusetts: Sinauer Associates.

Steinhardt, R. A., and Epel, D. (1974). Activation of sea urchin eggs by a calcium ionophore. *Proc. Natl. Acad. Sci. USA* **71,** 1915–1919.

NOTES TO THE PREPARATOR

1. Live sea urchins are available year-round from many suppliers. For a spring course, *Strongylocentrotus purpuratus* is appropriate because its breeding season goes from December through April. This sea urchin may be obtained from Pacific Bio-Marine Laboratories, Inc. (P.O. Box 536, Venice, California 90291). A "sea urchin kit" contains 50 urchins, KCl, plastic containers, droppers, directions, and filtered seawater. Air delivery by a specified date is possible. The kit is shipped on ice and needs to be stored at 2°C (maximum two days) until use. For a summer course, it may be necessary to use *Lytechinus pictus,* which breeds normally from May to September. This sea urchin may be purchased from Gulf Specimen Co., Inc. (P.O. Box 237, Panacea, Florida 32346).

2. To prepare sea urchins for spawning, inject each sea urchin with 1–2 ml of 0.5 M KCl into the body cavity with a 25-gauge needle inserted through the soft integument surrounding the oral opening. Within 5–15 min, males will produce a white exudate ("dry sperm") which should be allowed to flow directly, undiluted, onto a dry, sterile petri dish. "Dry sperm" may be stored at 4°C for up to 24 hr. Female urchins will produce a yellow or orange exudate that should be collected in a sterile plastic vessel containing a small volume of sea water. Eggs can be stored at 4°C for up to 6 hr.

3. Calcium ionophore is available from many supply companies, such as Sigma Chemical Company (P.O. Box 14508, St. Louis, Missouri 63178-9916).

NOTES

Early Development in the Sea Urchin and Starfish

As echinoderms, the starfish and the sea urchin have many characteristics in common. One of these is *embryogenesis,* or the manner in which embryos form and assume their typical anatomy.

The first few cleavages of these embryos are equal, or nearly so (Figure 2.1). By the 16-cell stage, however, the *blastomeres* (embryonic cells) are of three different sizes: the *mesomeres,* destined to produce most of the larval ectoderm; the *macromeres,* which give rise to the endoderm; and the *micromeres,* which develop into mesenchymal cells (a subset of the mesoderm). The mesenchyme, in turn, produces the larval skeleton consisting of calcareous spicules. The remainder of the mesoderm arises from the coelomic pouches that pinch off from the invaginated vegetal plate, or archenteron.

The *blastula* is a hollow ball of some 1000–2000 cells enclosing a fluid-filled cavity, the *blastocoel.* After hatching, the blastula settles on its ventral (vegetal-pole) side, which flattens into a *vegetal plate.* The micromere descendants of the plate migrate individually into the blastocoel, where they preferentially localize in the vegetal-pole region of the blastocoel. These are the *primary mesenchyme* cells. The macromere descendants of the plate invaginate into the blastocoel, forming a deep, narrow pocket, the *archenteron* (primitive gut). As the archenteron grows into the blastocoel, cells on its leading end (*secondary mesenchyme*) send out *filopodia* that contact the far (animal-pole) wall of the original blastula and establish connections with the interior of the animal-pole cells. This contact produces the anterior opening of the gut, or the *mouth.* (Where, how, and when did the *anus* form?)

At the end of gastrulation, the embryo consists of an outer cell layer (*ectoderm*), which grows a band or tuft of *cilia* for locomotion, and an inner cell layer formed by the invagi-

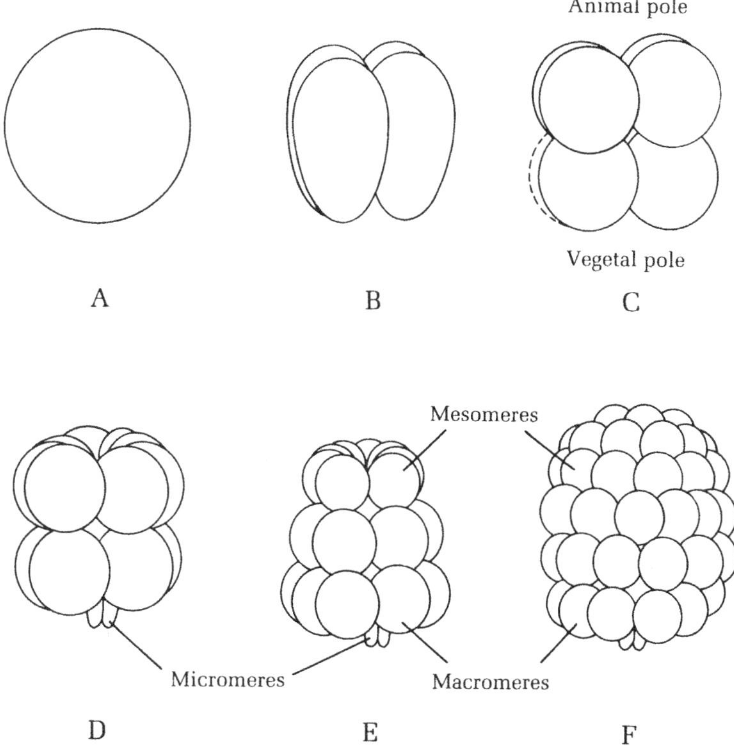

FIGURE 2.1

During their early developmental stages, the sea urchin and starfish look very much alike (A–I). During archenteron formation (J), however, the sea urchin larva acquires a conical shape (K); in contrast, the starfish bipinnaria is bean-shaped (L). (*Figure continues*)

nated vegetal plate. As the archenteron grows into the blastocoel, parts of its wall form outpocketings called *coelomic pouches*. These expand to line the blastocoel. The cavity enclosed by the expanded coelomic pouches is the *coelom*.

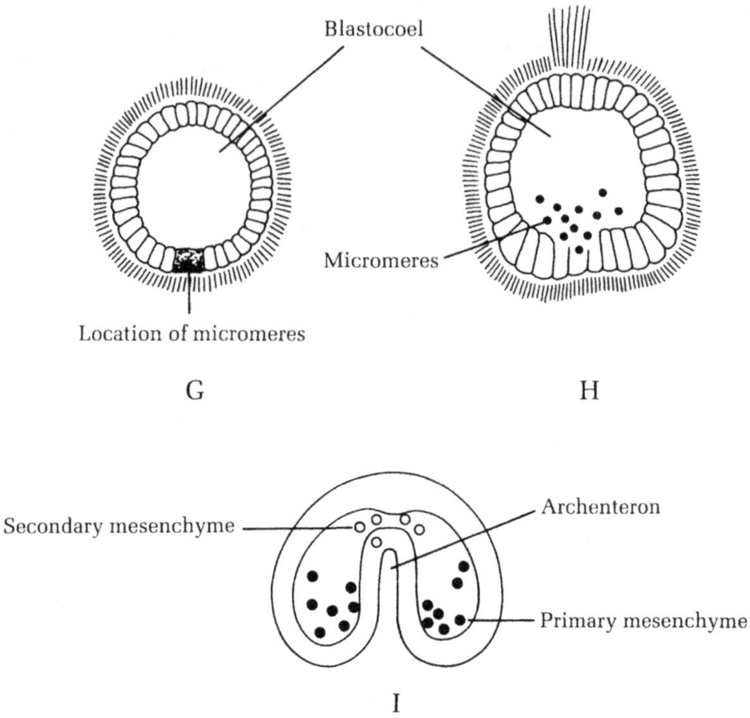

FIGURE 2.1 (continued)

Clearly then, the vegetal plate is the origin of the *endoderm* (gut) and of the *mesoderm* (mesenchyme cells and coelomic pouches). Therefore, the ectoderm is derived from the animal-pole cells, or those cells which in the blastula were not part of the vegetal plate. The sea urchin embryo develops

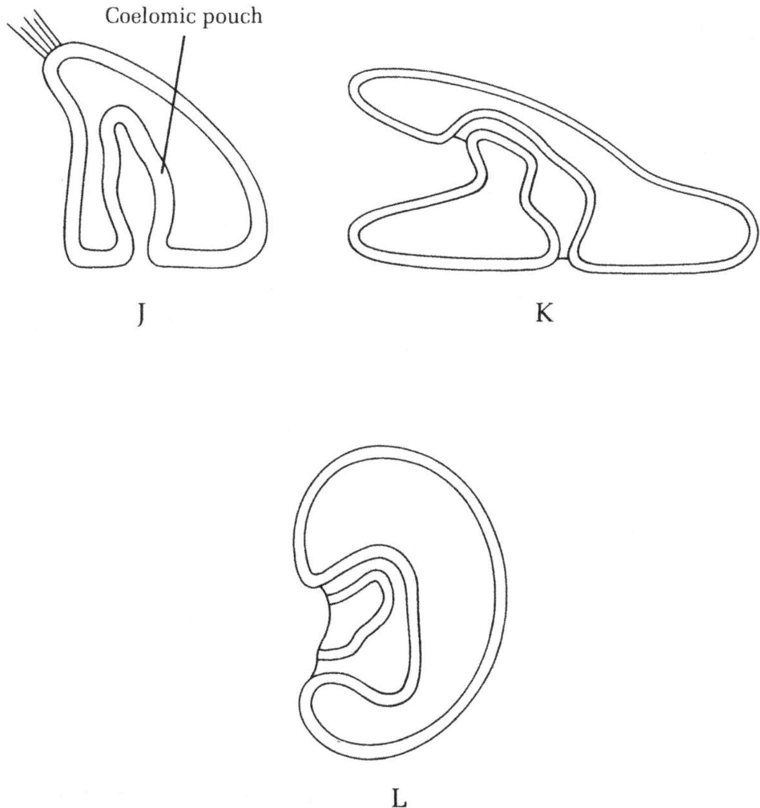

FIGURE 2.1 (continued)

into a larva called a *pluteus.* The corresponding larva in the starfish is called a *bipinnaria.* A drawing of sea urchin and starfish embryonic stages is included in this exercise (Figure 2.1).

PROCEDURE

1. Carefully compare each stage with the one before it and the one after it. Using the prepared slides of starfish and sea urchin embryos in your slide box, draw in your note pages representative embryos and label the parts indicated in Figure 2.1. Use black ink to ensure that attention has been paid to detail before the drawings are actually made. Indicate the magnification (Mag) of each drawing using the following formula. Examine any embryos you may have from the last lab period. To what stage have they progressed?

$$\text{Mag} = \frac{\text{Drawing}}{\text{Specimen}} = \frac{\text{Diameter of cell in drawing, mm}}{\substack{\text{Diameter of cell measured with} \\ \text{calibrated reticle, mm}}}$$

Example: $\dfrac{8.5 \text{ mm}}{0.04 \text{ mm}} = 212.5\times$ (may be rounded off to $212\times$)

2. The microscope reticle must be precalibrated with a stage micrometer. Use the reticle divisions to measure the actual size of your specimen, keeping in mind that each objective will impose a different magnification on your specimen. With each objective lens therefore, the reticle will have to be recalibrated.

3. The class will view a film or projection slides showing the different stages in sea urchin development. How does the coelom form? What other animals undergo similar coelom formation?

QUESTIONS

1. How does the gut form in the echinoderm embryo? Which orifice (oral or anal) forms first?

2. Assume that a mutation (dominant lethal) is discovered in sea urchins that uniformly kills embryos before the pluteus stage. On examination of these embryos, you realize they do not have spicules. What germ layer is affected by the mutation? Which blastomeres would you expect to be absent, defective, or abnormal in embryos with this mutation?

3. What is the fate of the primary mesenchyme cells? The secondary mesenchyme cells?

4. Calculate the volume of a blastomere in the one-, two-, and four-cell embryo. How do these volumes compare with that of a macromere at the 32-cell stage?

References

Balinsky, B. I. (1975). *An Introduction to Embryology,* 5th ed, pp. 135–143. Philadelphia: W. B. Saunders.

Gilbert, S. F. (1991). *Developmental Biology,* 3rd ed, pp. 114–115. Sunderland, Massachusetts: Sinauer Associates.

Giudice, G. (1973). *Developmental Biology of the Sea Urchin Embryo,* pp. 9–21. New York: Academic Press.

Hopper, A. F., and Hart, N. H. (1985). *Foundations of Animal Development,* 2nd ed, pp. 111–121. New York: Oxford University Press.

Mathews, W. W. (1986). *Atlas of Descriptive Embryology,* 4th ed, pp. 21–33. New York and London: Macmillan Publishing Co.

NOTES TO THE PREPARATOR

1. Prepared microscope slides of different echinoderm embryos are available from such biological supply companies as Ward's Natural Science Establishment, Inc. (5100 West Henrietta Road, P.O. Box 92912, Rochester, New York 14692-9012) and Carolina Biological Supply Company (2700 York Road, Burlington, North Carolina 27215).

2. If students were successful in growing embryos to the pluteus stage from the previous lab exercise, they should be encouraged to note down how plutei swim with only cilia. Provide dissecting microscopes and sterile plastic petri dishes with seawater for this purpose. Plutei may be transferred from one container to another with Pasteur pipettes fitted with rubber bulbs.

3. Films and projection slides are available from Pennsylvania State University (Audio-Visual Services, Special Services Building, University Park, Pennsylvania 16802), Biology Media (P.O. Box 10205, Berkeley, California 94709), Ward's Natural Science Establishment, Inc. (address given in Note 1), and Carolina Biological Supply Company (address given in Note 1).

NOTES

NOTES

Early Development in the Frog: To Neurulation

The unfertilized frog egg has two recognizable regions visible through the transparent *jelly coat:* the pigmented (dark gray), upper *animal hemisphere* and the yolky (yellow), lower *vegetal hemisphere.* During fertilization, the successful sperm enters the egg through the *sperm entry point,* a randomly determined point just above the egg's "equator." Simultaneously, the *gray crescent* (in certain species) appears directly opposite the sperm entry point.

Soon after fertilization, the egg undergoes its early *cleavages* (Figure 3.1). The first two cleavages are *equal,* yielding two and four *blastomeres,* respectively. Thereafter, cleavages are *unequal;* resumption of equal mitoses occurs later. Vegetal-pole cleavages occur more slowly and yield larger, yolky cells. When the embryo is several hundred cells old (not *big* — the cells become smaller with each division until after this stage), two distinct size classes of cells are present: the smaller, more numerous animal-pole cells and the larger, less numerous vegetal-pole cells. While these cleavages occur, the resulting cells become arranged into a hollow sphere, called a *blastula,* enclosing a fluid-filled cavity, the *blastocoel.*

Returning to the gray crescent, immediately beneath where the gray crescent was, an invaginated wrinkle (not unlike a pucker) forms, allowing the once exterior animal-pole cells in this area to sink into the blastocoel collectively, as a sheet. The sinking sheet progressively works its way in, its movement fueled by the rapid addition of newly formed surface cells. As the sheet sinks, two things happen: it encloses a cavity (the *archenteron* or "primitive gut") that gradually displaces the blastocoel, and the original site of the involution comes to define the *dorsal lip of the blastopore,* which presages the *anal opening* of the presumptive embryo. The first animal-pole cells to involute (the blind anterior end of the archenteron) proceed to establish contact with the inner surface of the once-hollow blastula. This results in the formation of the anterior opening of the forming embryonic gut, the *mouth* or *oral opening.* This entire process of involution (sinking of cells) and epiboly (movement of cell sheet over the embryonic surface) constitutes *gastrulation* (Figure 3.2).

As a result of gastrulation, the complement of cells that produces the *endoderm* is set aside. The endoderm thus arises from the involuted sheet of cells surrounding the archenteron and ultimately gives rise to the future gut. The cells that remain on the embryonic surface constitute the *ectoderm,* the fate of which will be described shortly. There is a third embryonic layer, the *mesoderm,* which first appears at the dorsal edge of the sinking cell sheet that later becomes the archenteron. This dorsal edge is the dorsal lip of the blastopore; its lateral and ventral lips subsequently form as gastrulation proceeds. The mesoderm gives rise to the muscles, bone, blood, connective tissues, and certain visceral organs of the future embryo.

At this point, you should study Figures 3.1 and 3.2 if you have not already done so. Review the preceding description until you can reconstruct these events mentally, without notes.

After gastrulation, the *ectoderm* has encircled the endodermal precursors that had earlier involuted. Between the endoderm and ectoderm, the mesoderm is growing rapidly. Eventually, a dense cord (or cylinder) of mesodermal cells, called the *chordamesoderm,* induces the overlying ectoderm on the future dorsal side of the embryo to form a pair of ridges, the *neural folds,* visible on the embryonic surface. The chordamesoderm eventually coalesces into the *notochord.* The ridges of the neural folds, meanwhile, grow higher and eventually coalesce, leaving a lumen or cavity running the length of the ridges. This is the *neural tube,*

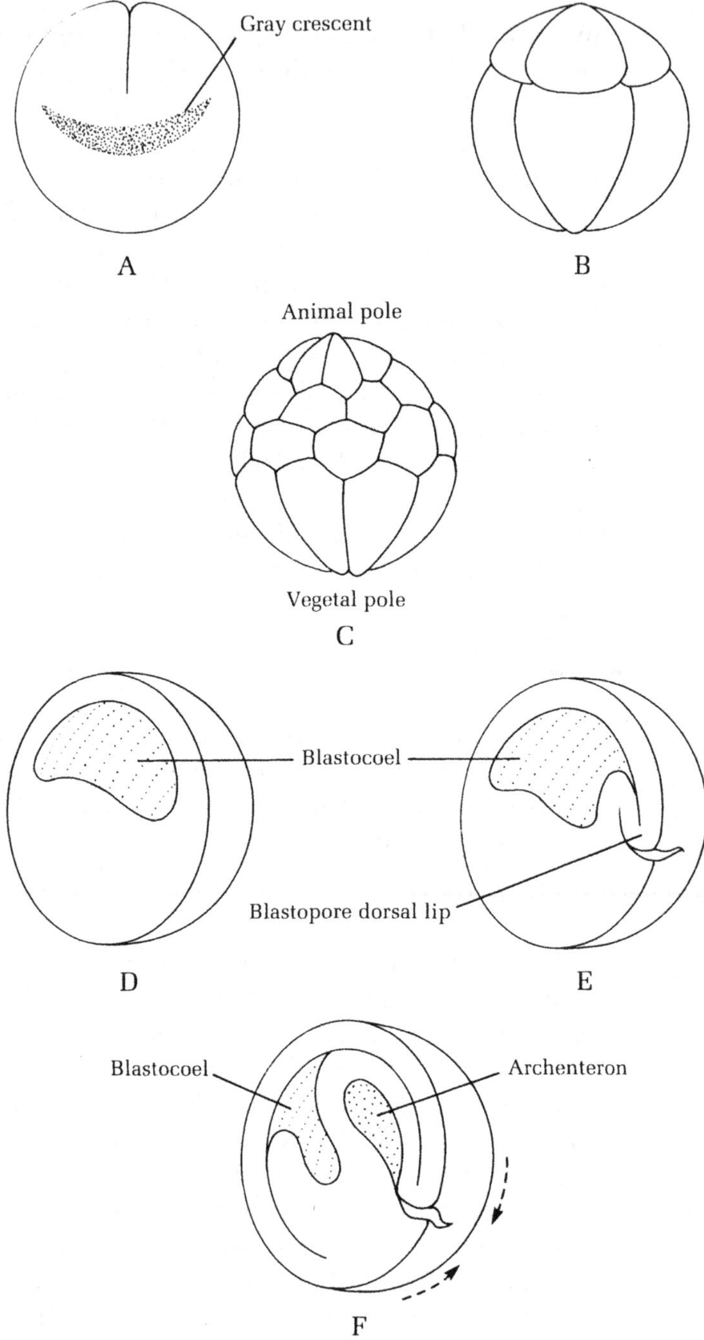
Gray crescent
A
B
Animal pole
Vegetal pole
C
Blastocoel
Blastocoel
Blastopore dorsal lip
D
E
Blastocoel
Archenteron
F

which will differentiate into the brain and spinal cord of the presumptive embryo which, at this stage, is called a *neurula*. This process is *neurulation.*

A few facts are worth pondering at this point. The ectoderm produces the skin (hair, nails, claws, and horns) of the developing embryo *and* also its brain, spinal cord, and nerves. In addition, as the ectodermal cells form the neural ridges, many of them pull away from the ectoderm, sink into the extracellular matrix of the mesodermal cells, become excluded from the forming neural tube, and *migrate* throughout the embryo, producing several cell populations, including pigment cells and the peripheral nervous system. These remarkable cells, the *neural crest cells,* have been the subject of intense research in developmental biology in recent years.

By now you should have a reasonably clear idea of the embryonic origin of the various organs and tissues, at least in vertebrates (the amphibia in particular). As we go on to study other embryos, you should compare how the embryonic germ layers (ectoderm, mesoderm, and endoderm) are generated.

FIGURE 3.1

Early cleavages and gastrulation in the amphibian embryo. (A) The plane of the first cleavage bisects the gray crescent (when present). (B) The third cleavage plane is subequatorial, unlike the first two, which are meridional. (C) The animal-pole cells divide more rapidly than the vegetal-pole cells. (D) Longitudinal section of the blastula showing the eccentrically positioned blastocoel. (E) Gastrulation commences with the formation of the blastopore dorsal lip. (F) Gastrulation creates the archenteron, which displaces the blastocoel. The arrows indicate the direction of invagination and involution of cells at the blastopore.

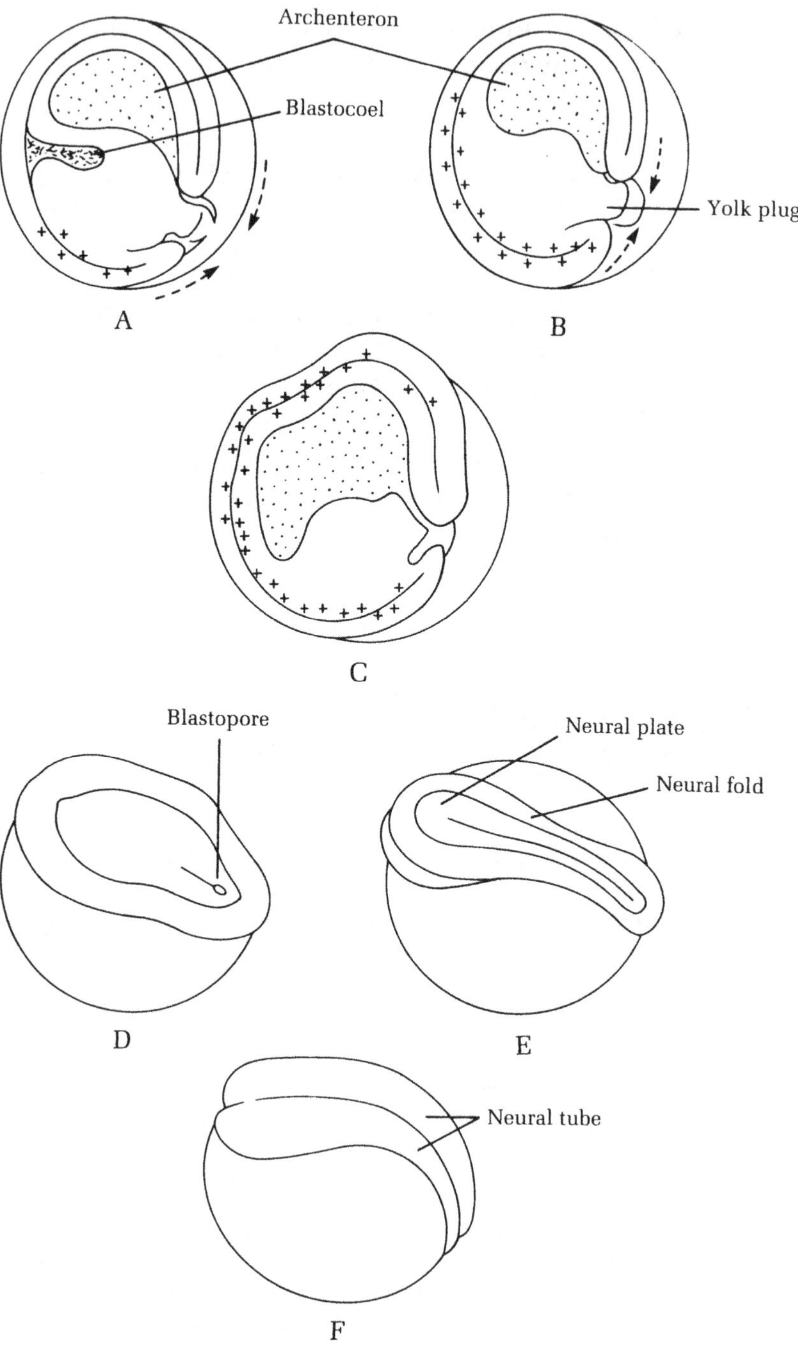
Archenteron
Blastocoel
Yolk plug
A
B
C
Blastopore
Neural plate
Neural fold
D
E
Neural tube
F

PROCEDURE

This lab exercise examines slide specimens consisting of sections of frog embryos (check your slide box), and three-dimensional plastic models of frog gastrulation and neurulation. The class should view a film or projection slides of early frog development, from fertilization to neurulation. Make drawings in the note pages of the various stages shown. Label your drawings (refer to Figures 3.1 and 3.2). Take down notes and concentrate on the cell movements that characterize early embryogenesis. This is often tedious (but not overly difficult), and requires a fair amount of concentration. It is often helpful to study with a partner and ask questions that may seem simple at the time.

FIGURE 3.2

Gastrulation and neurulation in the amphibian embryo. (A) The lateral and ventral lips of the blastopore eventually form, transiently capturing a portion of the invaginating yolky vegetal-pole cells which are visible externally as the yolk plug. (B) The blastocoel has been obliterated by the archenteron. (C) The yolk plug has receded completely into the archenteron; the blastopore eventually becomes the anus. Flattening of the exterior dorsal aspect of the embryo signals that neurulation has begun. (A–C) Crosses indicate mesoderm induction; the stippled areas are the archenteron; the shaded area is the disappearing blastocoel. (D) During the early stages of neurulation, the neural folds approach each other dorsomedially. (E and F) The neural plate sinks, becoming a neural groove, which comes to lie interiorly and becomes the cavity of the neural tube.

QUESTIONS

As you view the slides, make sure you consider the following questions and include in your drawings anatomical details that provide answers to these questions.

1. What is the normal position of the gray crescent with respect to the first cleavage furrow?

2. Describe the relative decrease in blastomere sizes (a) between the animal and vegetal pole at any one time, and (b) among blastomeres of the animal pole from the second cleavage to the blastula stage.

3. Describe the location of the pigment granules in the blastomeres. Are pigment granules present in the vegetal blastomeres?

4. Find bottle cells and illustrate several of them. What does their shape indicate about the cell movements occurring at this time?

5. Has the ventral blastopore lip formed prior to the total obliteration of the blastocoel?

6. Exactly where does the mesoderm first appear? How would you recognize these areas in the slide-mounted slices of gastrulas?

7. When is the notochord first apparent? How are its cells distinctive from those of the neighboring tissues?

8. What is the fate of the vegetal blastomeres? Estimate the fraction of these cells that go on to constitute the archenteron wall. Remember that the wall of the mature gut consists of an endodermally derived epithelial lining, mesodermally derived circular and longitudinal muscles, and mesodermally derived connective tissues.

References

Balinsky, B. I. (1975). *An Introduction to Embryology,* 5th ed, pp. 161–175. Philadelphia: W. B. Saunders.

Gilbert, S. F. (1991). *Developmental Biology,* 3rd ed, pp. 81–113, 121–138. Sunderland, Massachusetts: Sinauer Associates.

Hopper, A. F., and Hart, N. H. (1985). *Foundations of Animal Development,* 2nd ed, pp. 111–130. New York: Oxford University Press.

Nakamura, O., and Kishiyama, K. (1971). Prospective fates of blastomeres at the 32-cell stage of *Xenopus laevis* embryos. *Proc. Japan Acad.* **47**, 401–412.

Slack, J. M. W. (1983). *From Egg to Embryo,* pp. 31–65. Cambridge, England: Cambridge University Press.

NOTES TO THE PREPARATOR

1. Prepared microscope slides and plastic models of *Xeno-pus laevis* or *Rana pipens* embryos are available from many biological supply companies, such as Ward's Natural Science Establishment, Inc. (5100 West Henrietta Road, P.O. Box 92912, Rochester, New York 14692-9012) and Carolina Biological Supply Company (2700 York Road, Burlington, North Carolina 27215).

2. Films and projection slides of amphibian embryogenesis are available from the following major suppliers: Biology Media (P.O. Box 10205, Berkeley, California 94709), Kalmia Co., Inc. (71 Dudley Street, Cambridge, Massachusetts 02140), Pennsylvania State University (Audio Visual Sevices, Special Services Building, University Park, Pennsylvania 16802), and Carolina Biological Supply Company (address given in Note 1).

NOTES

NOTES

Early Development in the Frog: To 10 Millimeters

The three germ layers — endoderm, mesoderm, and ecto-derm — are in place by the time a frog embryo becomes a neurula. Rapid cell division and massive tissue rearrange-ments occur subsequently during *organogenesis,* a complex process that is not completely understood even today.

During *organogenesis,* the body of the embryo elongates (increases in length in the anteroposterior direction) and the tail forms. The body becomes subdivided into head and trunk, and the appendages develop. These massive changes have been chronicled for several organisms; the classic em-bryological models are the amphibians, birds, and fishes. A few mammals have been studied, but because these animals implant into maternal tissue during development, their em-bryos are much less accessible to study. Mouse and rat em-bryos are the most completely understood mammalian em-bryos; rabbit, monkey, and human embryos are less known.

This lab exercise involves microscopic examination of se-rial sections of early tadpoles (Figures 4.1, 4.2). A serial sec-tion is composed of consecutive slices (longitudinal or *along* the anterior–posterior axis of the body; sagittal if the section cuts right through the median anterior–posterior axis of the animal; cross or transverse, if the section cuts *across* the an-terior–posterior axis of the body). When looking at the slides of the frog embryos, orient yourself by locating the first slice, then gradually work, slice by slice, through the specimen. Imagine looking at a slow-motion film through the animal, following anatomical features to see how they are positioned with respect to each other.

Figures 4.1 and 4.2 will guide you through the specimens using anatomical landmarks. You are expected to be able to identify the parts labeled in these drawings. The *surest* way to remember details is by making your own drawings. In the note pages, draw at least *one representative* (meaning fairly complex) slice from each slide. It may be necessary to use the dissecting microscope for the larger slices. Indicate magnifications.

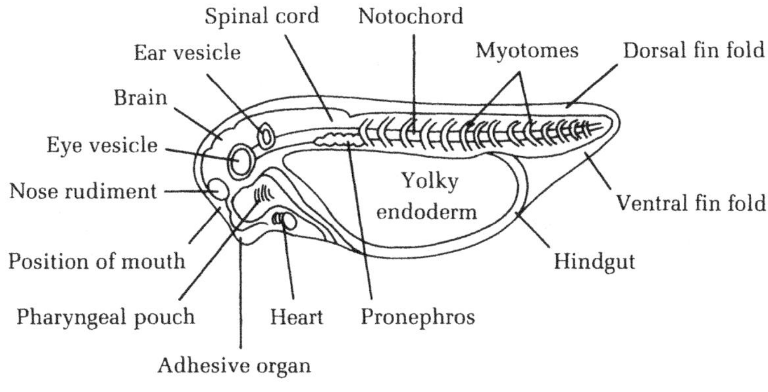

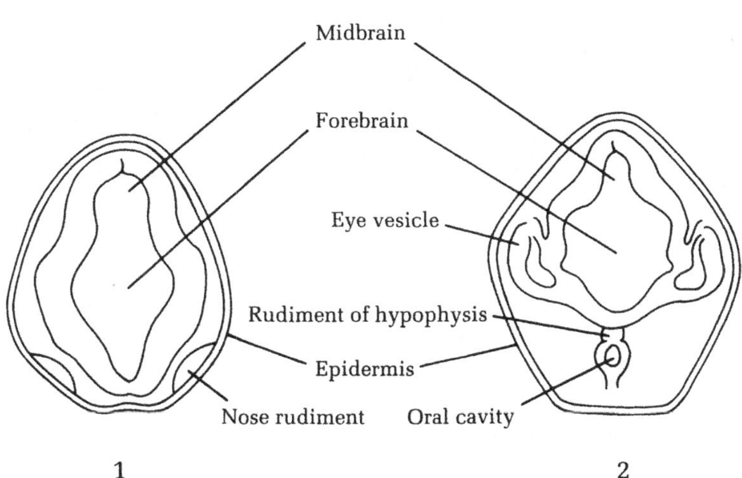

FIGURE 4.1

Longitudinal and cross-sectional views of a 4-mm frog embryo. Numbered lines represent the planes corresponding to the numbered transverse sections.(*Figure continues*)

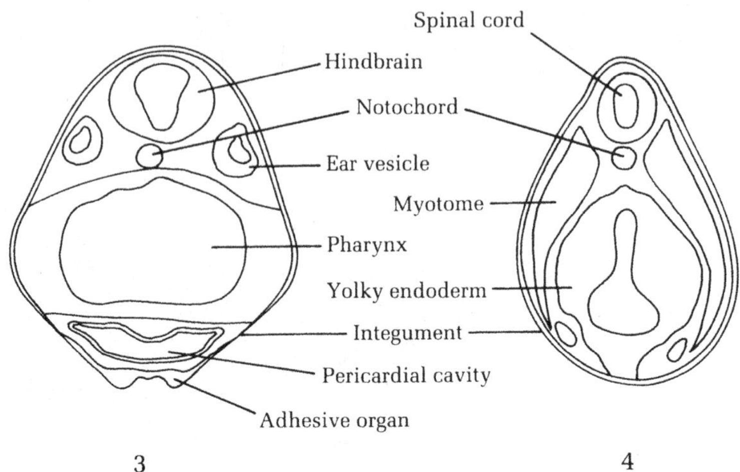

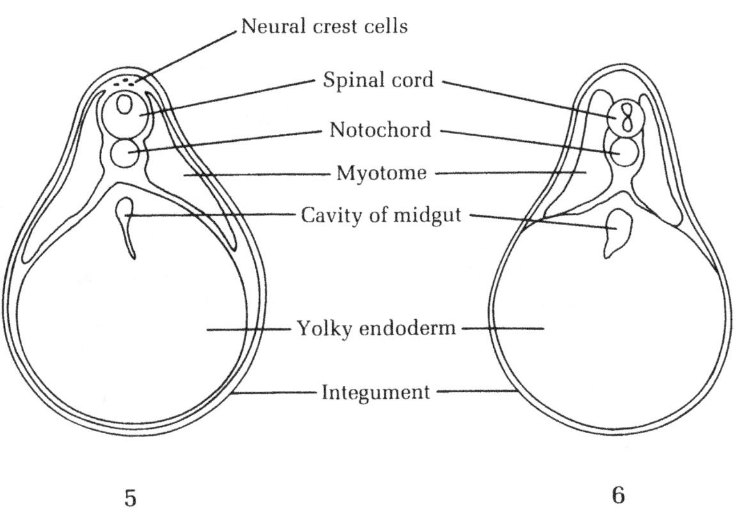

FIGURE 4.1 (continued)

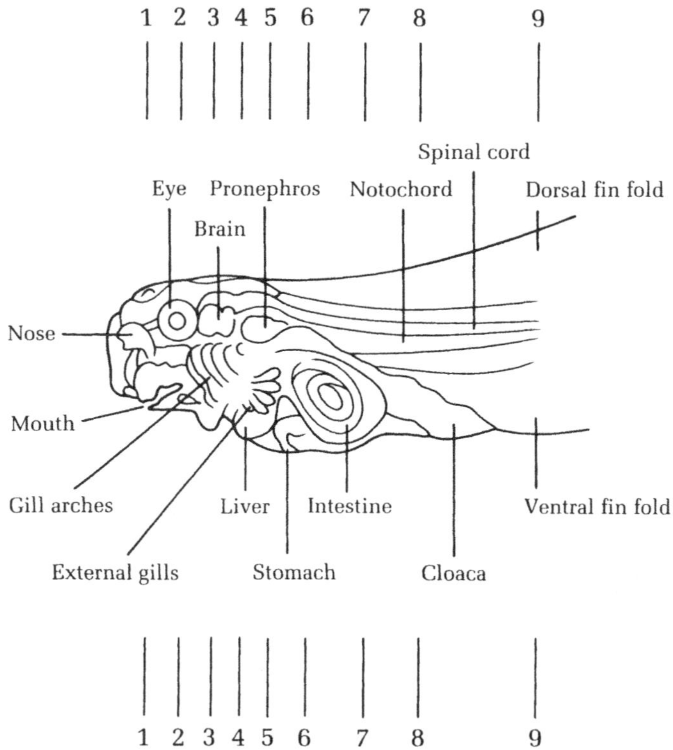

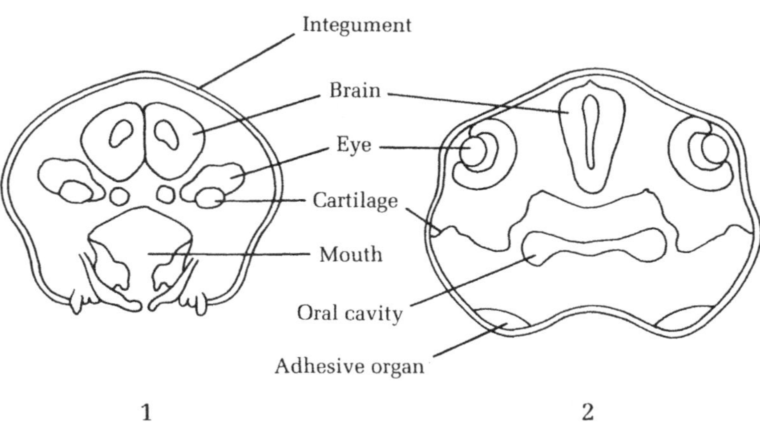

FIGURE 4.2

Longitudinal and cross-sectional views of a 10-mm frog embryo. Numbered lines represent the planes corresponding to the numbered transverse sections.

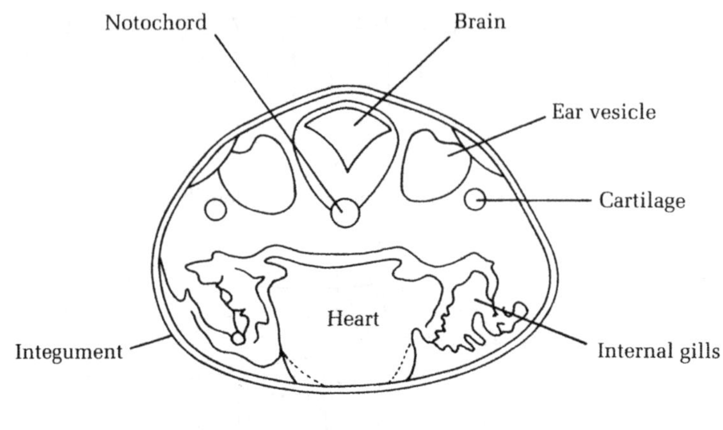

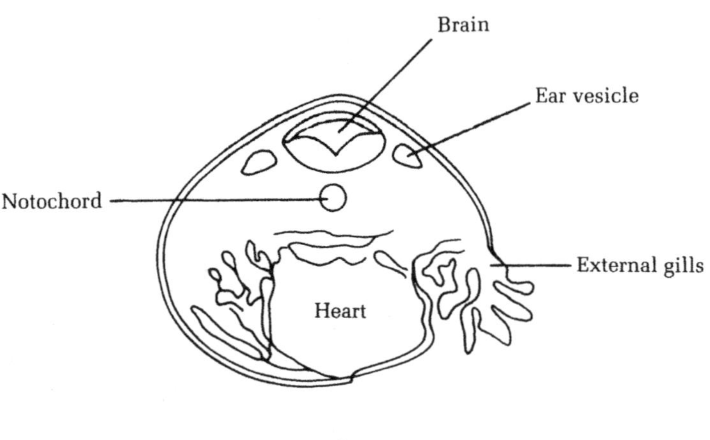

FIGURE 4.2 (continued)

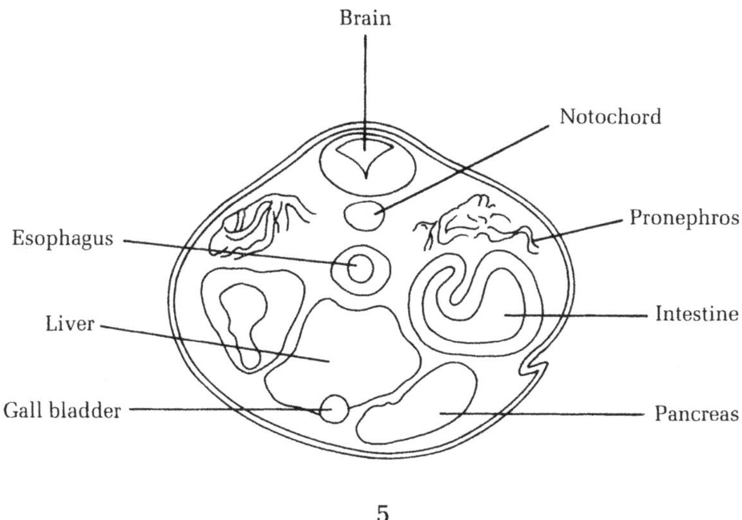

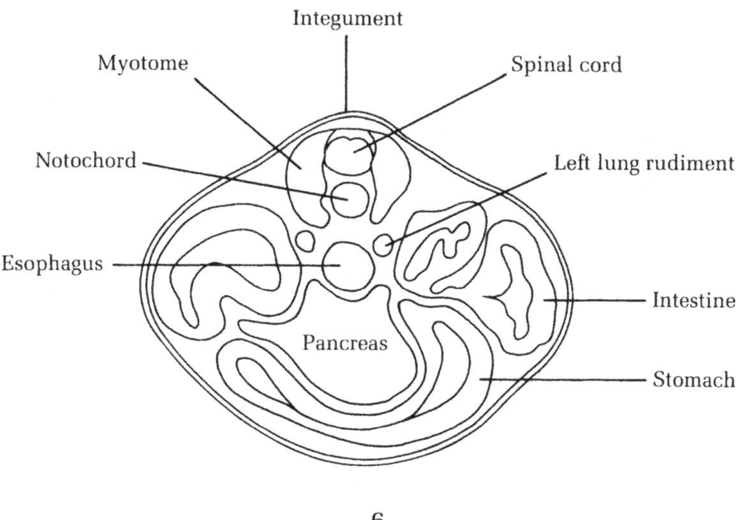

FIGURE 4.2 (continued)

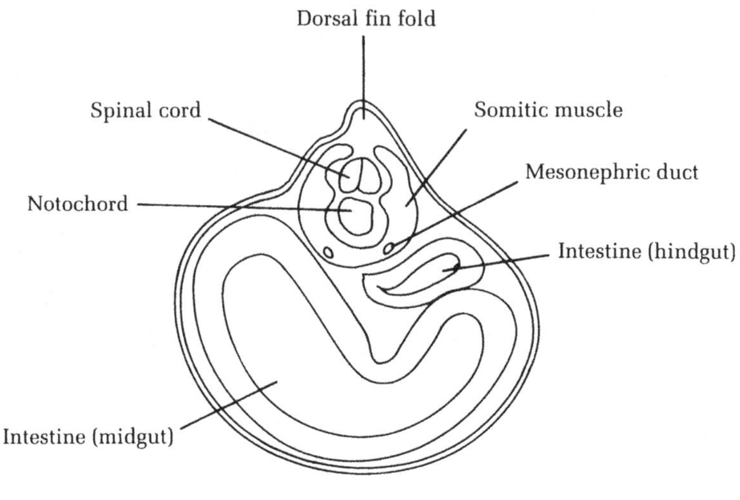

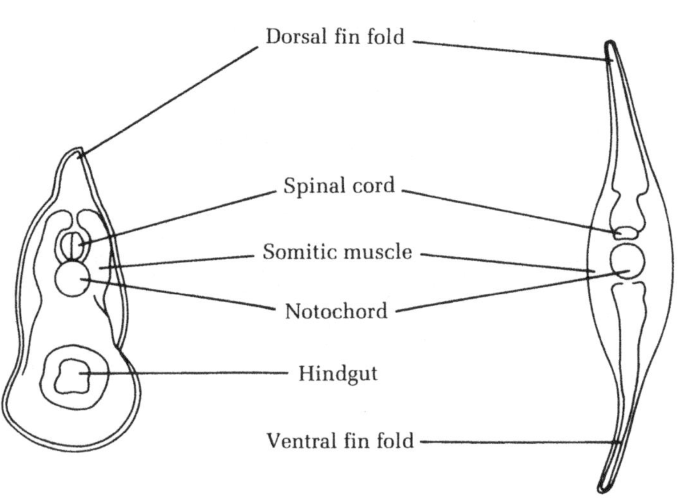

FIGURE 4.2 (continued)

PROCEDURE

1. Locate the slides for this exercise in the slide box. Take them all out (2.5-mm, 4-mm, 5- to 7-mm, and 10-mm embryos) and superficially examine them in order to have a feel for what the slides are portraying. It is best to locate the starting section first and then move the slide sequentially through the other sections.

2. When changing objective lenses on the microscope, be careful not to hit the slide. The slide itself is expensive, but the objective lens is much more so. An oil immersion **objective** although not normally necessary, may be used if needed.

3. The work is tedious but patience and effort will reward you with a much better feel for morphogenesis in general. It is *not* a good idea to put off the hours of work required in this exercise.

QUESTIONS

1. Locate the notochord and chart its development. Does it grow larger or longer? This structure is the common denominator humans share with all other chordates, but it does not persist in all chordates. What is the developmental significance of the notochord?

2. From what germ layer does the lens of the eye originate? What about the liver, skin, lungs, kidney, and gut? To answer these questions, review the slides of younger embryos and locate in them the primordia of these organs.

3. Recall the blastopore in the frog gastrula and neurula. What structure does it correspond to now? Justify your answer.

4. When do the lungs first appear? Why not earlier? Be careful with your answer. Remember that (a) function follows form, and (b) evolution is a process that has no preconceptions, grand design, or "purpose." It simply has consequences and proximate causes.

5. What is the embryonic origin of the pituitary gland? Explain.

6. What are the differences between dermatome, myotome, and sclerotome?

7. What is the relationship between the pharyngeal pouches and the external gills in the tadpole?

8. What is the cloaca? In which other chordates is the cloaca present?

References

Eichler, V. B. (1978). *Atlas of Comparative Embryology*, pp. 46–67. St. Louis: The C. V. Mosby Co.

Gilbert, S. F. (1991). *Developmental Biology,* 3rd ed, pp. 157–159. Sunderland, Massachusetts: Sinauer Associates.

Mathews, W. W. (1986). *Atlas of Descriptive Embryology,* 4th ed, pp. 54–102. New York: Macmillan Publishing Co.

Oppenheimer, S. B., and Chao, R. L. C. (1984). *Atlas of Embryonic Development,* pp. 14–53. Newton, Massachusetts: Allyn and Bacon.

Slack, J. M. W. (1983). *From Egg to Embryo,* pp. 31–65. Cambridge, England: Cambridge University Press.

Wischnitzer, S. (1975). *Atlas and Laboratory Guide for Vertebrate Embryology,* pp. 56–122. New York: McGraw-Hill Book Co.

NOTE TO THE PREPARATOR

Prepared microscope slides and plastic models of *Xenopus laevis* or *Rana pipiens* embryos are available from many biological supply companies, such as Ward's Natural Science Establishment, Inc. (5100 West Henrietta Road, P.O. Box 92912, Rochester, New York 14692-9012) and Carolina Biological Supply Company (2700 York Road, Burlington, North Carolina 27215).

NOTES

NOTES

Preventing Axis Formation in Amphibians with Ultraviolet Irradiation

Following fertilization of a frog egg, the cortex of the egg cytoplasm immediately shifts relative to the inner cytoplasmic depths, resulting in a rapid rearrangement of the gray pigment granules on the surface of the animal hemisphere. The blanched area from which the pigment granules emigrate is the *gray crescent.* Beneath the gray crescent, the *blastopore* soon forms. Internally, this event is equivalent to establishing the anteroposterior axis. the blastopore marking the posterior or caudal end of the forming embryo.

Cytoplasmic rearrangements are crucial to axis formation in amphibian embryos (Figure 5.1). If these rearrangments are prevented from taking place on schedule, the resulting embryo is "ventralized," or all-belly: to use Spemann's term, *Bauchstucke.* Such an embryo lacks a body axis and dorsal structures. Recall that in the amphibian embryo, the vegetal pole is engulfed by the animal pole during gastrulation; across the animal pole, the dorsal axis is established, with the blastopore position defining the caudal end of the embryo.

The application of ultraviolet (UV) light to the vegetal pole of a frog embryo has been shown to cause *Bauchstucke* formation. This result was originally interpreted to be evidence for the existence of dorsal determinants in the cytoplasmic cortex. "Determinants" in this sense refers to specific particles or substances, the presence of which directs axis formation. (Recall the "organizer" theory of Spemann and coworkers as discussed in lecture.) When these determinants are disrupted or disabled by UV irradiation, they cease to promote axis formation. This simple and elegant means of explaining axis formation requires only that determinants be present in their normal form; if either distribution or form is altered, no axis develops.

More recent experiments have shown that the presence of such special substances need not be invoked. Rather, the formation of *Bauchstucke* embryos probably has a simpler ex-

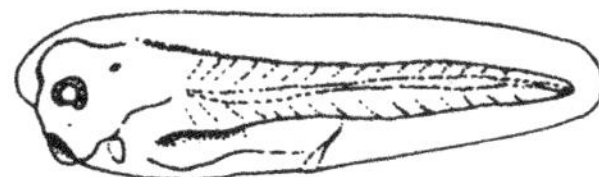

DAI 5: Normal tadpole with one adhesive gland, two eyes, two otic vesicles. The vesicles are identifiable by little white otoconia.

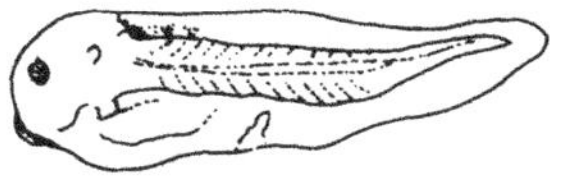

DAI 4: Slightly microcephalic. Forehead reduced. Both eyes present, but may be small.

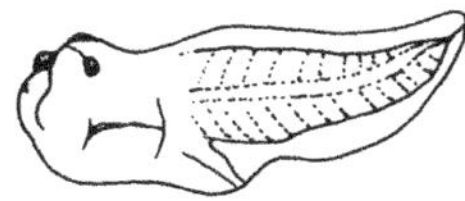

DAI 3: Microcephalic. Cyclopic, but eye pigment (black) present

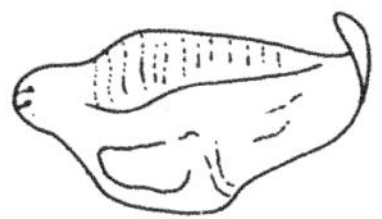

DAI 2: Extremely microcephalic. Some head structures present. No eye pigment. Otic vesicles present. Adhesive gland (medium brown) present.

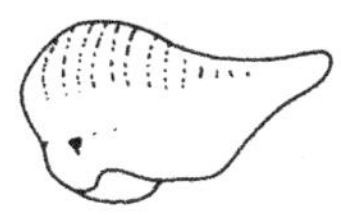

DAI 1: Acephalic (no head structures present). No otic vesicles or adhesive gland. Somites present.

DAI 0: Aneural (neural folds never develop). No somites. Consists of ciliated epidermis, primitive gut, mesenchyme, and blood. Equivalent to Bauchstucke of Spemann.

FIGURE 5.1

Dorsoanterior (DAI) Index Spectrum (redrawn from Kao and Elinson 1988). The DAI is a value that describes physical malformations in *Xenopus laevis* tadpoles exposed as embryos to ultraviolet irradiation. It appears that exposure to ultraviolet rays foreshortens the anterior region and truncates the development of dorsal structures. The most severely malformed tadpoles are assigned a DAI value of 0; normal tadpoles are given a DAI value of 5.

planation: UV light disrupts the cytoplasmic rearrangements (approximately a 30° rotation) that occur in the normal course of development. Without these rearrangements, secondary induction fails, and thus, no mesoderm forms. This interpretation is borne out by other observations: UV-irradiated eggs inclined obliquely and briefly to gravity develop into normal tadpoles. In this instance, exposure to gravity brings about cytoplasmic rearrangements that compensate for UV-disrupted rearrangements. Therefore, UV irradiation itself disrupts no actual substances or particles acting as axis determinants. Axis determination, then, is the collective result of cytoplasmic rearrangments, and is not narrowly attributable to the mere presence of specific substances or molecules.

This lab exercise will test the hypothesis that cytoplasmic rearrangements are necessary for normal development of the amphibian embryonic axis, evaluating embryos irradiated with UV light that are subsequently "rescued" by gravity. If UV irradiation has physically altered the determinant (that is, if we get *Bauchstucke* embryos), then all embryos, regardless of subsequent rescue by gravity, should be *Bauchstucke*. However, if UV-irradiated but gravity-rescued tadpoles have a normal body axis, then the function of the determinant is determined not by its mere presence, but also by its distribution in the cytoplasm and a need for its activation which UV irradiation disrupts.

The most important thing to keep in mind for this lab exercise is that you will have to be ready when the embryos are. Therefore, your time is not your own. Prepare for the actual experiments by reading this procedure several times and making a dry run. Gravid frogs, available in the lab, are to be stripped of their eggs only when you are completely ready. The eggs need to be fertilized as soon as sperm can be obtained from explanted, macerated testes. The instructor should assist with minor calamities but will not take credit for any successes that are not hers. This leaves the student with an immense responsibility. Because the class is small and the exercise demands intense concentration, students should work individually. In this way, the student can rightfully claim credit for any successes.

PROCEDURE

1. Prepare the immobilizing device first. This is an agar dish with little holes melted out of the agar. The dishes have already been prepared. To make the holes, hold the tip of a Pasteur pipette to a flame (candle or Bunsen burner) so that a little ball of glass forms at the pipette tip. This ball should be about 1 mm in diameter in order to melt holes in the agar into which *Xenopus* eggs will fit snugly.

2. Melt 4 groups of 6 holes each in the agar. Fill the dish with 5% Ficoll solution (dissolved in 20% Ringer's solution) and then pour off all the Ficoll until only a thin film remains on the agar surface. Use a pipette or a pair of forceps to remove the bubbles in the holes. (You could make each group size bigger than 6 eggs, but it is more manageable to repeat the experiment in order to get more data than to do a lot of embryos all at once.)

3. Fertilize and dejelly *Xenopus* eggs. Obtain *Xenopus* eggs by stripping a gravid female. This will be demonstrated. View the eggs under a dissecting microscope so that you can identify the interface between the vitelline envelope and the jelly coat.

4. Prepare sperm suspension. This also will be demonstrated. Use a clean dropper to spread the sperm, in drops, over the eggs.

5. Immediately view the eggs under the dissecting microscope. Soon after eggs are fertilized, their jelly coats delaminate from the egg surface, appearing eventually as halos around the eggs. You should be able to see this happen. When no more fertilization envelopes form, it is time to move to the next step.

You should have read this material in advance, because at this point you should know that fertilized eggs need to be irradiated as soon as possible — so do not waste any time.

6. Dejelly the eggs by rinsing them in cysteine hydrochloride solution. This reduces the disulfide bonds that give the jelly its firm structure. As soon as the jelly disappears, rinse the eggs in two good swirls of Ringer's solution.

7. Secure enough eggs to make the following groups:

   A. Those to be irradiated for 30 sec, and within this group:

      a. Those to be positioned animal-pole up

      b. Those to be positioned equator up

   B. Those to be irradiated for 45 sec, and within this group:

      a. Those to be positioned animal-pole up

      b. Those to be positioned equator up

   C. Those not to be irradiated (leave in dish)

8. Transfer irradiated eggs to the agar dish into which Ficoll has been poured. Using a smooth glass probe, or a clean pair of CLOSED and somewhat blunt forceps, and your ingenuity, position the egg on the edge of the hole and gently push or roll it into the hole. Thus, for Group Aa, for example, set the egg on the "precipice" animal-pole up, then rotate it into the hole so that its equator bobs to the top as the egg fits snugly into the hole. This is not easy. To rattle you further, keep in mind that you have to record the time when these positionings were done.

9. Keep all eggs in their immobilized position until those that are animal-pole up have cleaved. Note any changes in pigmentation. Again, record the time.

10. Finally, for each group, remove all immobilized eggs from their holes by aspirating them with an appropriately sized Pasteur pipette. Transfer all the eggs in one group (keep the groups separate) into a dish with Ringer's solution in it.

11. Observe the next several cleavages. For reference, a timetable of normal embryonic development is included in this exercise. Incubate the embryos and keep recording your observations. Remove any embryos that are clearly dead (shriveled, flattened, or disintegrated). Before discarding these embryos, however, *make sure that any anatomical abnormalities are duly noted according to the Dorsoanterior Index Spectrum* (Figure 5.1). Change the Ringer's solution if necessary.

12. Your results will be presented to the class during the next lab and fully discussed in a written report to be submitted later.

QUESTIONS

1. Use short sentences and diagrams to summarize what you did. Explain your diagrams as you would to a student (with the same qualifications as you) about to do the same experiment. State your hypothesis, premises, expectations, and anticipated problems. Write up this experiment as you did the sea urchin fertilization lab exercise.

2. Suggest techniques that you have learned from performing this exercise that might be useful for next year's class.

References

Gilbert, S. F. (1991). *Developmental Biology*, 3rd ed, pp. 305–308. Sunderland Massachusetts: Sinauer Associates.

Gruss, P., and Kessel, M. (1991). Axial specification in higher vertebrates. *Curr. Opinion Genet. Dev.* **1**, 204–210.

Kao, K. R., and Elinson, R. P. (1985). Alteration of the anterior–posterior embryonic axis: The pattern of gastrulation in macrocephalic frog embryos. *Dev. Biol.* **107**, 239–251.

Kao, K. R., and Elinson, R. P. (1988). The entire mesodermal mantle behaves as Spemann's organizer in dorsoanterior enhanced *Xenopus laevis* embryos. *Dev. Biol.* **127**, 64–77.

New, H. V., Howes, G. and Smith, J. C. (1991). Inductive interactions in early embryonic development. *Curr. Opinion Genet. Dev.* **1**, 196–203.

Scharf, S. R., and Gerhart, J. C. (1983). Axis determination in eggs of *Xenopus laevis:* A critical period before first cleavage, identified by the common effects of cold, pressure, and ultraviolet irradiation. *Dev. Biol.* **99**, 75–87.

Timetable of Normal Embryonic Development in *Xenopus laevis,* the South African Clawed Frog, at Room Temperature (22–24°C)

Stage	*Time*
1-celled	0
2-celled	1.5 hr
4-celled	2.0 hr
8-celled	2.25 hr
16-celled	2.75 hr
32-celled	3.0 hr
Blastula	5.0 hr
Gastrula	9.0 hr
Yolk plug	13.5 hr
Neural plate	16.25 hr
Neural fold	18.25 hr
Neural groove	19.75 hr
Neural tube	20.75 hr
Hatching	50 hr
Begin feeding	98 hr
Appearance of hind limbs	24 days
Appearance of fore limbs	44 days
Froglet	58 days

NOTES TO THE PREPARATOR

1. *Xenopus laevis* is used for this experiment. One gravid female can provide sufficient eggs for a class of 20. *Xenopus laevis* is available from most biological supply companies in the United States, such as Ward's Natural Science Establishment, Inc. (5100 West Henrietta Road, P.O. Box 92912, Rochester, New York 14692-9012) and Carolina Biological Supply Company (2700 York Road, Burlington, North Carolina 27215). However, use of these animals may be subject to state restrictions because they are not indigenous to North America. Frogs and tadpoles must be destroyed after use.

2. To obtain sufficient numbers of eggs at specified times, female frogs will have to be superovulated with two intraperitoneal injections, 36 hr apart, of 50 IU pregnant mare serum gonadotropin and 800 IU human chorionic gonadotropin. Both hormones, as well as reagents, are available from Sigma Chemical Company (P.O. Box 14508, St. Louis, Missouri 63178-9916). Female frogs should be kept separate from males.

3. Female frogs are stripped of eggs just as the experiment begins. The eggs are fertilized by adding drops of sperm suspension obtained by macerating testes dissected from male frogs. The testes are explanted whole and kept in a 40% modified Ringer's solution until used. A small piece, about one-third the size of the average testis, is macerated, with fine forceps or iridectomy scissors, in 1 ml of a 40% modified Ringer's solution:

    2.922 g NaCl

    0.956 g KCl

    0.095 g magnesium chloride ($MgCl_2$)

    0.147 g calcium chloride ($CaCl_2$)

    500 ml double distilled water

    Adjust pH to 7 with 2.5 mM HEPES buffer.

4. Fertilized eggs are dejellied in 2.5% cysteine hydrochloride in 40% modified Ringer's solution, adjusted to pH 7.8 with 1 *M* NaOH. Dejellied eggs are rinsed with a 20% modified Ringer's solution and immobilized in solidified 5% agar on which a thin film of 5% Ficoll in 20% modified Ringer's solution has been applied.

5. Embryos to be irradiated are placed in a UV-permeable cuvette and exposed to shortwave UV rays from a UV lamp (Scientific Products, Baxter Healthcare Corp., 1430 Waukegan Road, McGaw Park, Illinois 60085-6787).

6. Statistical analysis may be applied to pooled class data, if desired.

NOTES

NOTES

NOTES

Early Chick Development: To 48 Hours

Cleavage in the early embryo may be complete (holoblastic) or incomplete (meroblastic). Thus far, the embryos we have studied all undergo *holoblastic* cleavage, in which cytokinesis is rapidly completed following nuclear division.

In the chick embryo, cleavage is initially *meroblastic;* the cytoplasm cleaves incompletely and remains contiguous with the large mass of yolk that later provides nutrients to the developing embryo.

Unlike the frog egg, the chicken egg has an enormous amount of yolk packaged along with it. Because the yolk is heavier than non-yolky cytoplasm, the latter is displaced into a small flattened spot (the *blastodisc*) on the surface of the yolky mass. The blastodisc contains the united male and female nuclei, and as cleavage occurs, the nonyolky cytoplasm becomes incompletely cleaved into *blastomeres.* These blastomeres constitute the *blastoderm,* which defines the animal pole of the egg.

The external layer of cells of the blastoderm constitutes the *epiblast,* from which will be derived the ectoderm, mesoderm, and endoderm of the future embryo. The greatly flattened *blastocoel* separates the epiblast from a more interior, thin layer of cells, the *hypoblast,* which appears to be equivalent to the parietal region of the mammalian primitive endoderm. The hypoblast itself is separated from the yolk by the *subgerminal cavity.* The early cleavages occur during the egg's passage through the oviduct, prior to shell deposition.

By the time the chicken egg is laid, the circular, lens-shaped blastoderm consists of a dark outer border, the *area opaca,* and a translucent central area, the *area pellucida.* The area pellucida varies in thickness, its posterior half (the *embryonic shield*) being composed of a greater number of cells than the anterior half. The *hypoblast* originates from the embryonic shield.

The *primitive streak* is a thickening of epiblast cells that appears in the blastoderm and grows anteriorly, then moves posteriorly (originating from the embryonic shield area). This is the first obvious embryonic structure to develop in the chicken egg (see Figure 6.1), and it was named long before its role as an organizational center was appreciated. The appearance of the primitive streak marks the beginning of pregastrulation (14–17 hours after incubation). A fully developed primitive streak has a median furrow, the *primitive groove;* at the anterior (advancing) end of the streak is a broad circular thickening in the blastoderm, the *primitive knot* (more popularly known as *Hensen's node*).

The primitive streak elongates by the concentration of more and more cells from the periphery of the blastoderm toward the *primitive groove.* In the region of the groove, at Hensen's node, the surface cells of the blastoderm appear to sink singly (the epithelium does not "crawl" in the vicinity of the node) and disappear beneath the *epiblast.* These migrating cells establish contact with the underlying *hypoblast* cells and proceed to spread out sideways and forward from the anterior end of the primitive streak. Note that the primitive streak is not itself a fixed anatomical structure, but merely a formation that persists as a result of the continued ingression of surface cells. The *epiblast* is the upper, or outer, embryonic epithelium; the *hypoblast* is the lower, or inner, epithelium.

As the cells enter the primitive streak, the streak elongates toward the presumptive head region. The first cells to migrate through the streak are those destined to become the

At this junction it sometimes helps to remember the amphibian blastopore and its dorsal lip. Note how the frog blastopore and chick primitive streak are dynamic, constantly changing cell populations. Note, too, that the blastopore dorsal lip and Hensen's node appear to have the same "sink" function. There is a significant difference between these two systems, however. In amphibians, the cells move *collectively* in sheets; in the chick, the cells move *individually* as they sink beneath the blastoderm surface.

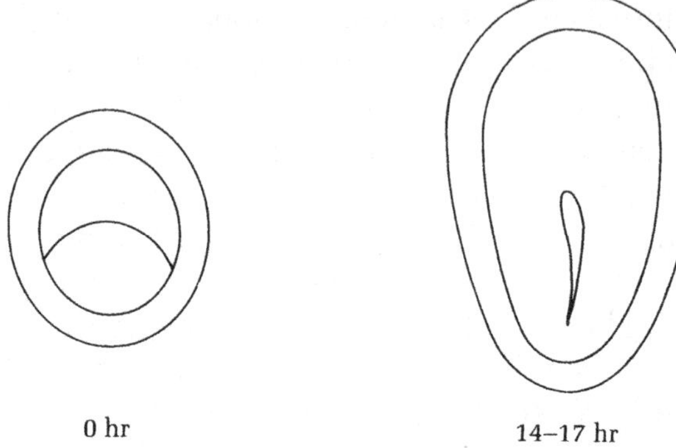

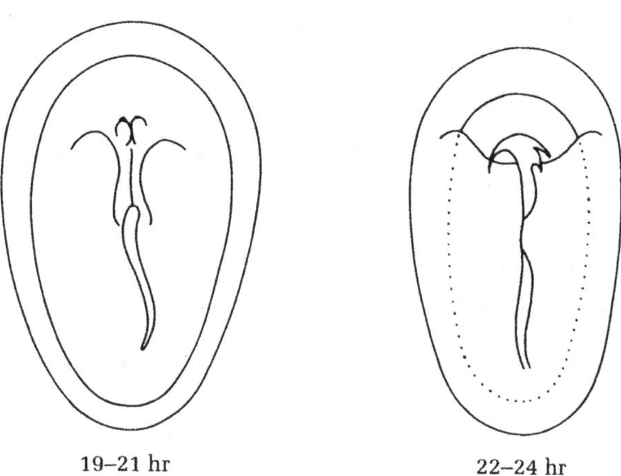

FIGURE 6.1

Stages of chick development.

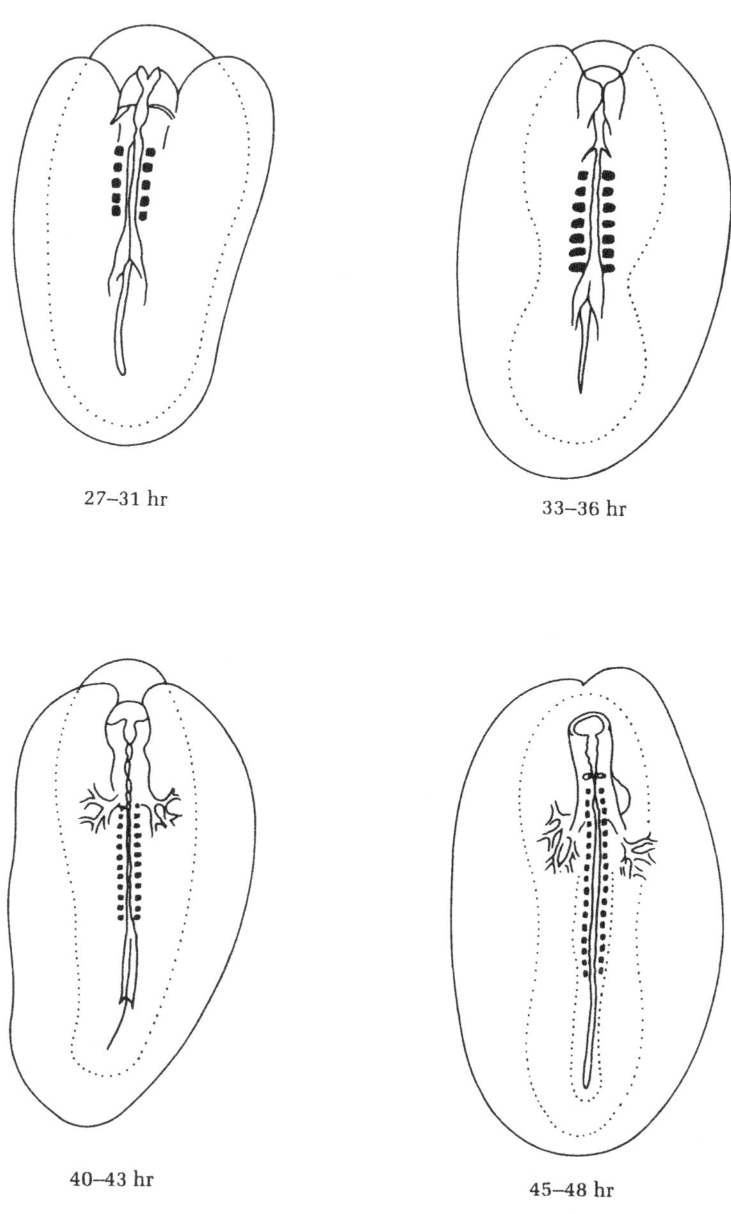

FIGURE 6.1 (continued)

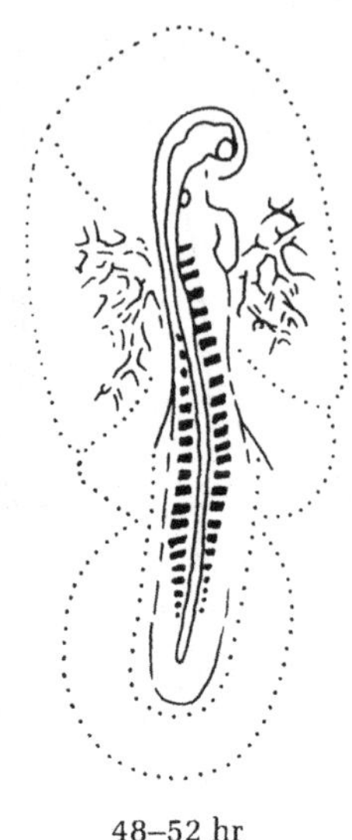

48–52 hr

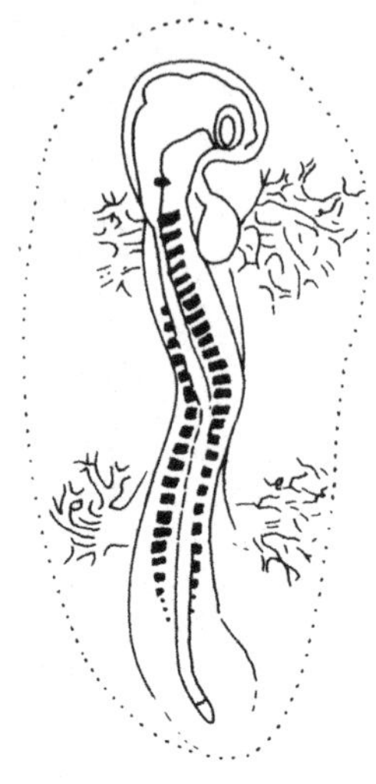

50–53 hr

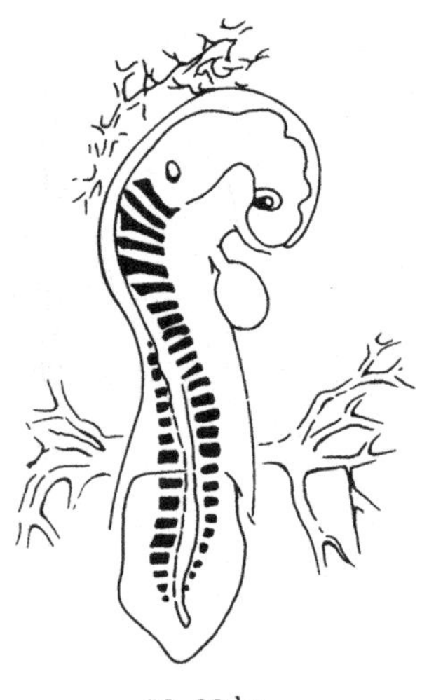

56–60 hr

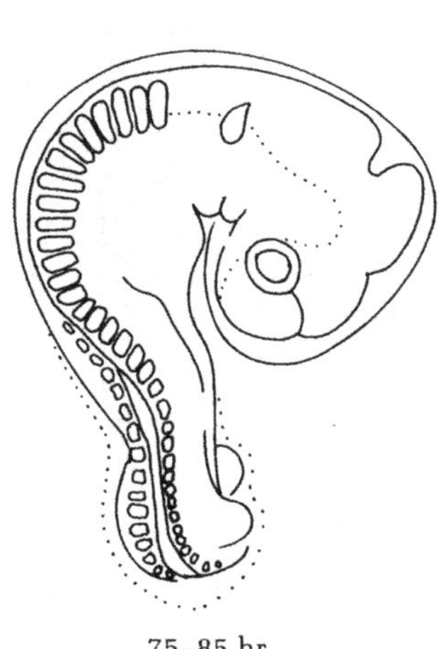

75–85 hr

FIGURE 6.1 (continued)

foregut (*endoderm*). The next group of cells to sink in through Hensen's node also moves anteriorly, remaining just above the first group and just beneath the epiblast to form the *head mesoderm* and the *notochordal cells.* Later, ingressing cells separate into two streams, one joining the hypoblast and producing, eventually, the remaining *endodermal tissues* of the embryo. The second stream of cells spreads through the blastocoel as a loose sheet and generates the *mesodermal portions* of the embryo. The *yolk sac* and *allantois* are formed from the hypoblast and from some of the second stream of cells. This ends the twenty-second hour of incubation. The process is not over yet. Hensen's node truly represents the blastopore, as will soon be explained.

While the mesodermal (second) stream of cells continues to move into the embryo, the primitive streak regresses, pulling Hensen's node posteriorly along the longitudinal axis of the early embryo. This regression leaves in its wake the *head process.* The node moves even farther posteriorly, laying the precursors of the notochord in its retreating wake. Finally, the node moves all the way most posteriorly, forming the anal region of the early embryo (Figure 6.2). (Remember the blastopore?) By this time the epiblast is composed entirely of ectodermal cells. (Why?) This sounds quick and easy, but the retreat of Hensen's node took 24–26 hr.

This lab exercise involves examining whole mounts and sections of chick embryos (transverse and sagittal) at the primitive streak, at 18-, 24-, 33-, and 48-hr stages. Do not panic. There *is* a way to organize all this into a coherent set of events. *First,* understand what is going on from the introductory lecture. If you do not understand something, ask — this is not the time to hesitate. *Then,* when you know what is going on, look at the whole mounts, youngest to oldest. *Finally,* examine the sections. This should remind you of the frog sections. In order to stay on course, do what ballet dancers do: spot. Fix your gaze at one structure (for example, the notochord, an obvious one) and locate that structure in every section you subsequently examine. This will give you an orientation.

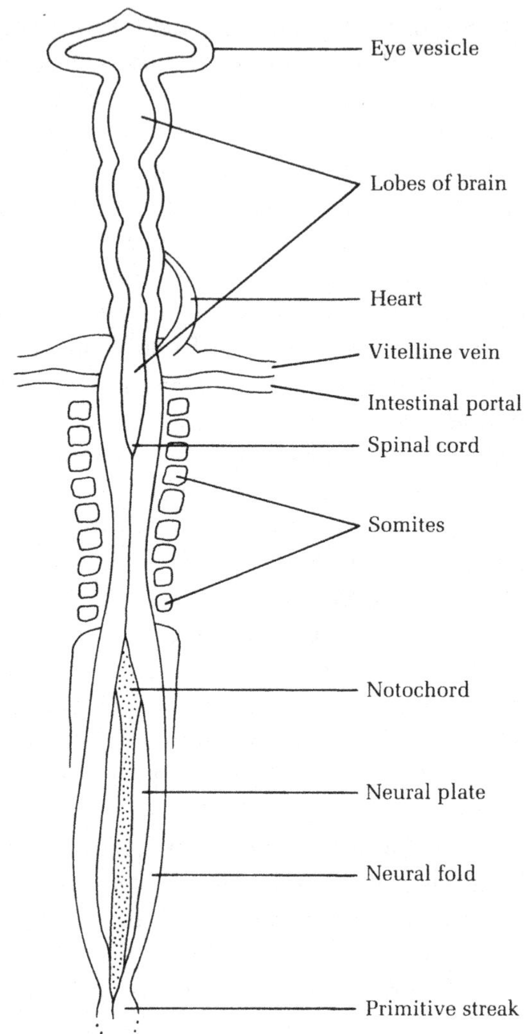

FIGURE 6.2

Chick embryo, 29 – 30 hr.

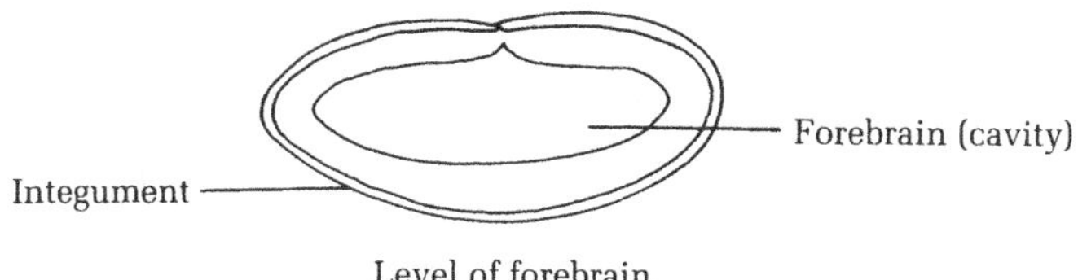

Level of forebrain

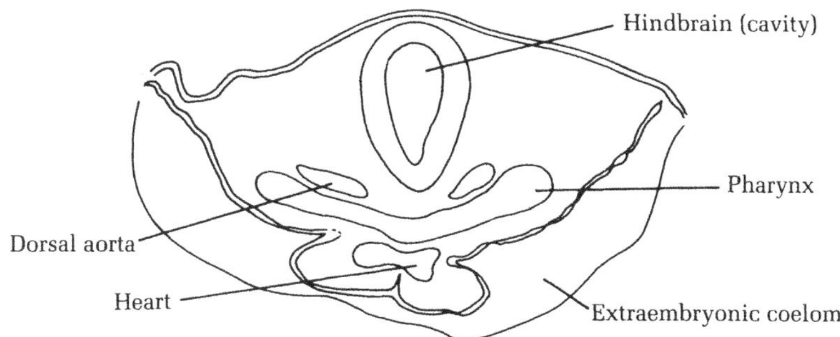

Level of hindbrain; the heart is single at this level,
but it clearly originates from a left and a right half.

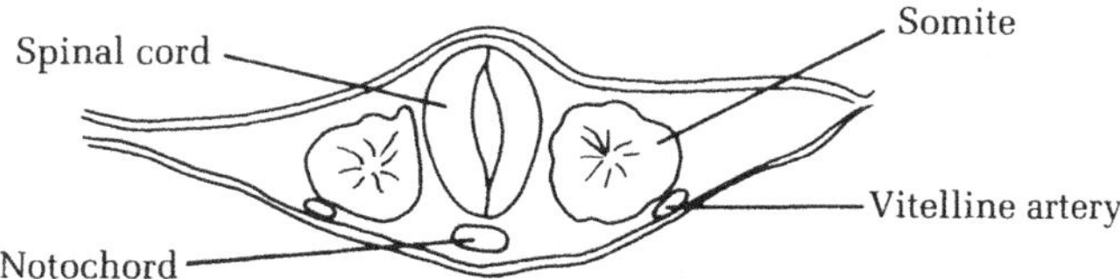

Level of trunk with closed neural tube.

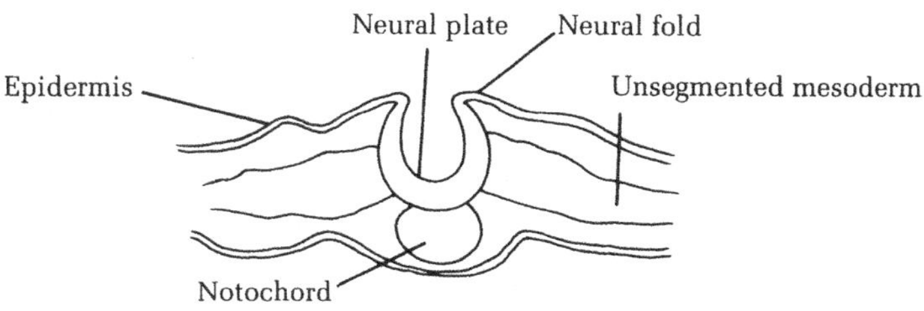

Level of trunk with open neural plate.

FIGURE 6.2 (continued)

PROCEDURE

Locate the structures labeled in Figures 6.1 and 6.2 in your specimens. Be sure you can recognize the major visceral organs. Making your own ink drawings in the note pages provided will greatly help in identifying and remembering structures. For the larger sections, use the dissecting microscope.

QUESTIONS

1. Does the chick embryo have an archenteron? If so, where does it come from and how does it form? If not, why not? (This is a profound question — do not take it lightly.)

2. Compare the frog blastopore and the chick primitive streak. Use these two structures to defend the classification of animals into Protostomia and Deuterostomia. Be prepared to elaborate on your answer during a class discussion.

3. The chick embryo is said to have an anteroposterior developmental gradient. This means that the anterior end is in a more advanced stage of development than the posterior end, which does not catch up until development is complete. With the initial and final positions of Hensen's node in mind, explain how the gradient arises.

4. This lab exercise closely ties in with the next exercise. If you feel you have located all the parts labeled in Figures 6.1 and 6.2, and have made your own drawings, you may proceed to the next exercise.

References

Eichler, V. B. (1978). *Atlas of Comparative Embryology*, pp. 76–81. St. Louis: The C. V. Mosby Co.

Gilbert, S. F. (1991). *Developmental Biology,* 3rd ed, pp. 138–146. Sunderland, Massachusetts: Sinauer Associates.

Mathews, W. W. (1986). *Atlas of Descriptive Embryology,* 4th ed, pp. 109–121. New York: Macmillan Publishing Co.

Newby, W. W. (1960). *A Guide to the Study of Development,* pp. 18–34. Philadelphia: W.B. Saunders Co.

Oppenheimer, S. B., and Chao, R. L. C. (1984). *Atlas of Embryonic Development,* pp. 57–63. Newton, Massachusetts: Allyn and Bacon, Inc.

Rugh, R. (1977). *A Guide to Vertebrate Development,* 7th ed, pp. 131–150. Minneapolis: Burgess Publishing Co.

Slack, J. M. W. (1983). *From Egg to Embryo,* pp. 152–161. Cambridge, England: Cambridge University Press.

Watterson, R. L., Schoenwolf, G. C., and Sweeney, R. M. (1979). *Laboratory Studies of Chick, Pig, and Frog Embryos,* 4th ed, pp. 4–32. Minneapolis: Burgess Publishing Co.

Wischnitzer, S. (1975). *Atlas and Laboratory Guide for Vertebrate Embryology,* pp. 40–57. New York: McGraw-Hill Book Co.

NOTE TO THE PREPARATOR

Prepared slides of sectioned and whole-mounted chick embryos, as well as plastic models, are available from most biological supply houses, such as Ward's Natural Science Establishment, Inc. (5100 West Henrietta Road, P.O. Box 92912, Rochester, New York 14692-9012) and Carolina Biological Supply Company (2700 York Road, Burlington, North Carolina 27215).

NOTES

NOTES

Early Chick Development: To 96 Hours

As the chick embryo develops, the curvature of the head and body becomes more pronounced. Eventually, the embryo comes to lie on its left side. Organogenesis is well underway, as is the formation of the extraembryonic tissues: *amnion, chorion, yolk sac,* and *allantois* (Figures 7.1, 7.2).

In this lab exercise, whole mounts and serial sections of 72- and 96-hour chick embryos will be examined. Most of these slides are fragile and should be handled carefully. For some of these slides, the dissecting microscope can provide sufficient magnification and will prove to be more useful than the compound microscope.

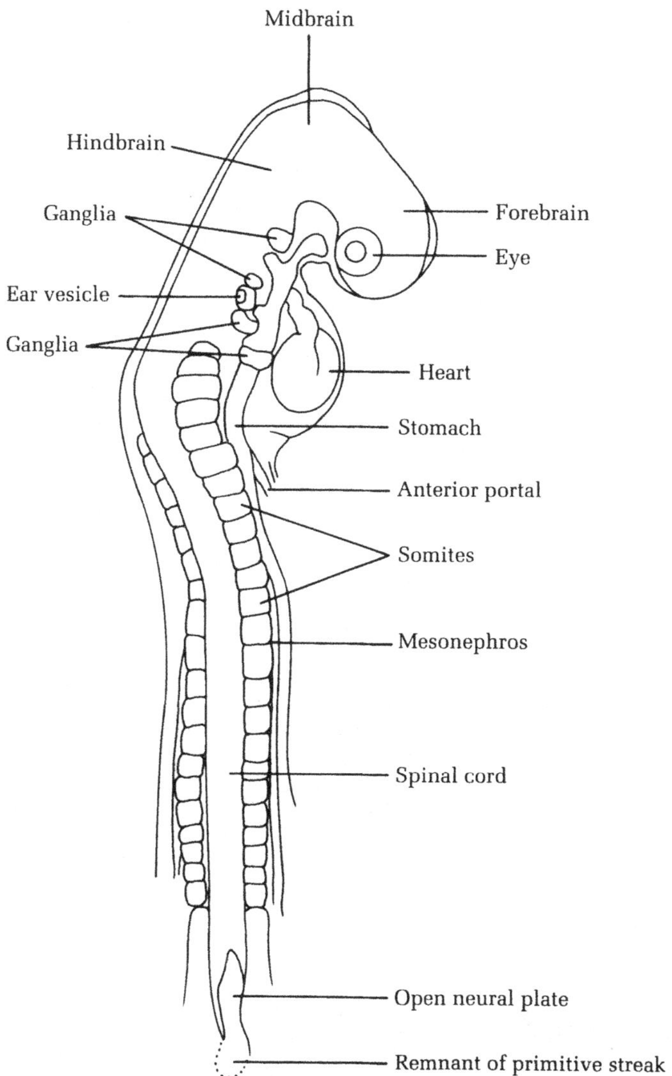

FIGURE 7.1

Chick embryos, 50 – 55 and 68 – 72 hr.

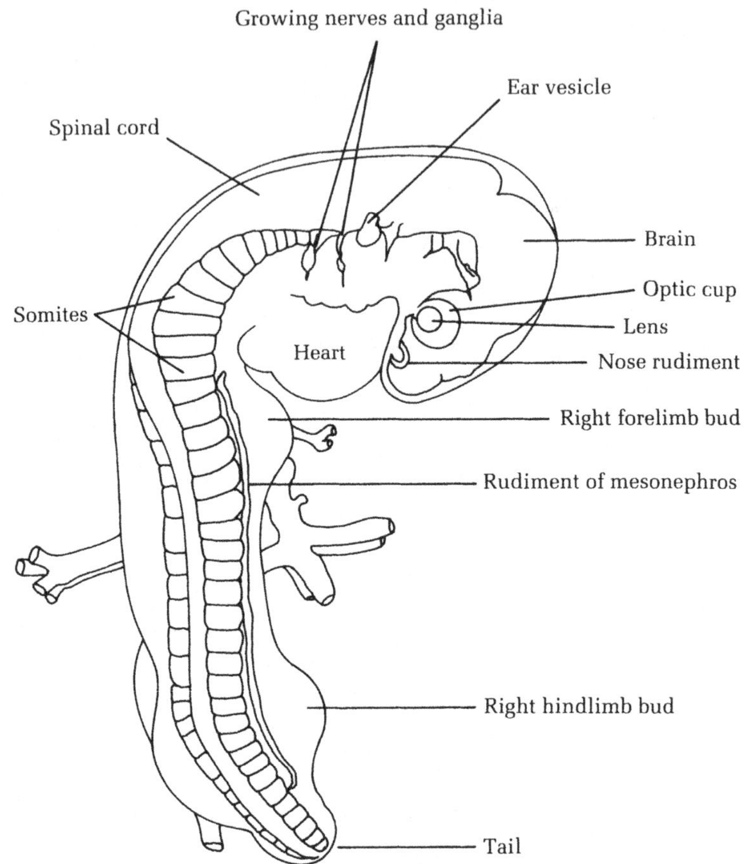

Chick embryo, 68–72 hr

FIGURE 7.1   *(continued)*

Midbrain
Ganglion
Notochord
Dorsal aorta
Pharynx
Mandibular arch
Head mesenchyme
Optic cup
Lens vesicle
Forebrain
Vitelline vessel
Level of the eye
Spinal cord
Somite
Descending aorta
Stomach
Sinus venosus
Ventricle
Vitelline vessel
Level of the stomach

Spinal cord
Somite
Notochord
Descending aorta
Stomach
Vitelline veins
Liver diverticula
Ventricle
Level of the liver
Foregut
Vitelline vein
Anterior intestinal portal
Vitelline vessel
Level of the intestinal portal

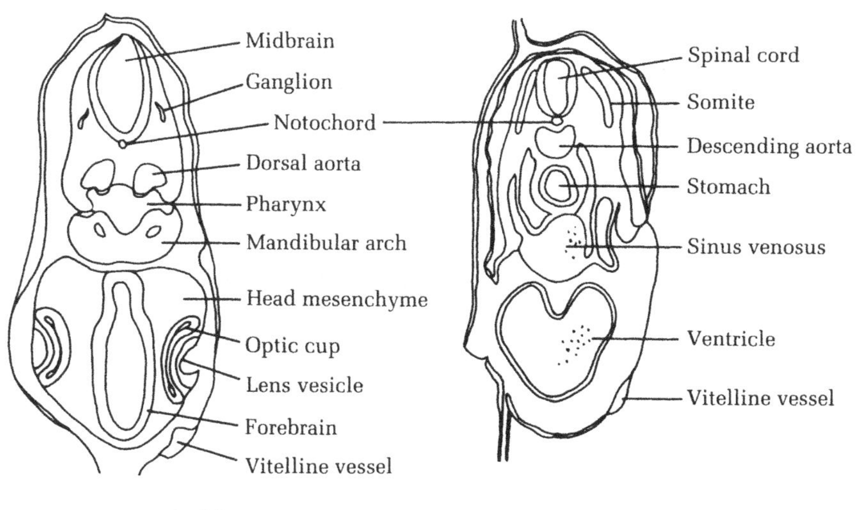

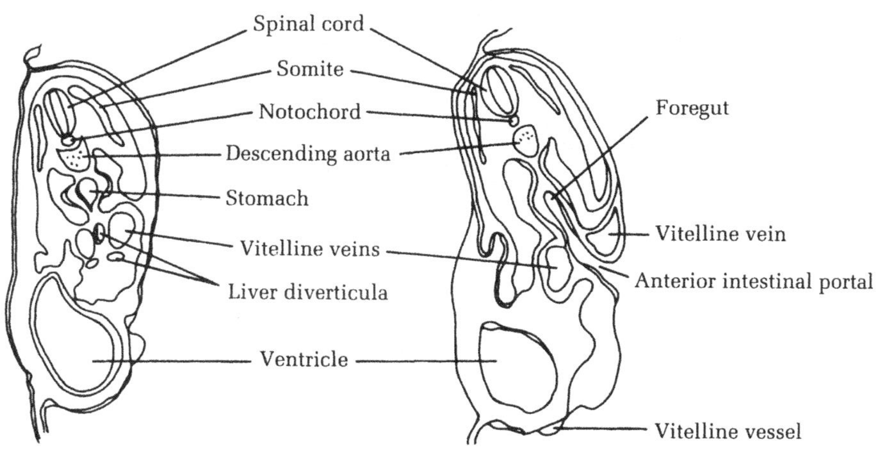

FIGURE 7.2

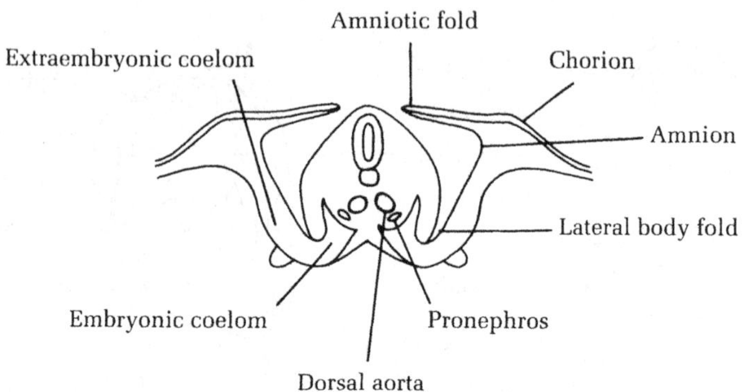

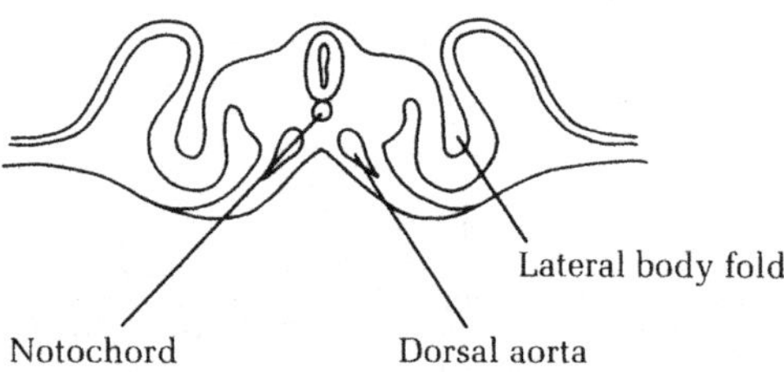

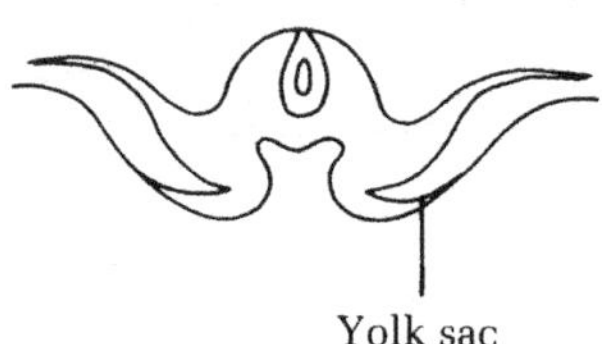

FIGURE 7.2   *(continued)*

PROCEDURE

Examine the specimens and make ink drawings in your notebook of representative whole embryos and sections. Figures 7.1 and 7.2 should help you identify the various body parts, internal organs, and extraembryonic tissues of the developing chick embryo. A film or projection slides will be shown demonstrating the development of chick embryos from fertilization to hatching. Use these images along with the microscope images to answer the following questions.

QUESTIONS

1. Describe how the spinal column forms around the spinal cord. What tissues are involved? At what developmental age do the first vertebrae appear?

2. How is a chondrocyte different from a mesenchymal cell or fibroblast?

3. In the oldest whole-mounted chick embryo available, how many somites are there? What is a somite, anyway? Describe its origin and fate.

4. What is the anterior portal? To what is it a "portal"?

5. Be sure to label the four extraembryonic tissues in your
    drawings. How do you know that your identification is
    accurate? What landmarks did you use?

6. Not all the extraembryonic tissues are vascular. Which
    are? Explain the functional significance of such vascu-
    larization.

References

Balinsky, B. I. (1975). *An Introduction to Embryology,* 4th ed, pp. 313–329. Philadelphia: W. B. Saunders Co.

Eichler, V. B. (1978). *Atlas of Comparative Embryology,* pp. 83–119. St Louis: The C.V. Mosby Co.

Mathews, W. W. (1986). *Atlas of Descriptive Embryology,* 4th ed, pp. 123–179. New York: Macmillan Publishing Co.

Newby, W. W. (1960). *A Guide to the Study of Development,* pp. 43–66. Philadelphia: W. B. Saunders Co.

Oppenheimer, S. B., and Chao, R. L. C. (1984). *Atlas of Embryonic Development,* pp. 65–103. Newton, Massachusetts: Allyn and Bacon, Inc.

Rugh, R. (1977). *A Guide to Vertebrate Development,* 7th ed, pp. 152–212. Minneapolis: Burgess Publishing Co.

Wischnitzer, S. (1975). *Atlas and Laboratory Guide for Vertebrate Embryology,* pp. 59–89. New York: McGraw-Hill Book Co.

NOTES TO THE PREPARATOR

1. Prepared slides of sectioned and whole-mount chick embryos are available from most biological supply companies, such as Ward's Natural Science Establishment, Inc. (5100 West Henrietta Road, P.O. Box 92912, Rochester, New York 14692-9012) and Carolina Biological Supply Company (2700 York Road, Burlington, North Carolina 27215).

2. Films and projection slides of developing chick embryos from blastodisc to hatching are available from Biology Media (P.O. Box 10205, Berkeley, California 94709) and Pennsylvania State University (Audio-Visual Services, Special Services Building, University Park, Pennsylvania 16802).

3. Many students are unfamiliar with avian internal gross anatomy and welcome the opportunity to examine such features as the nonfunctional right oviduct and ovary, internally located testes, gizzard, crop, and cloaca. For this purpose, procure a freshly dressed hen and rooster and demonstrate these anatomical specializations by dissecting a bird of each sex.

NOTES

NOTES

Chick Development in a Windowed Egg

The chicken (*Gallus gallus*) egg is a standard specimen in embryology and developmental biology. The embryo is relatively large and, more importantly, it is accessible during nearly all stages of development. (Eggs are laid during the later meroblastic cleavages—when fertile.)

A large body of basic information on chick development exists. The developmental stages have been standardized as "stage numbers" or "hours." Modern chick studies involve tissue grafting (chicken–quail chimeras) and cell migration studies (neural crest cells).

In this lab exercise, a fertile chicken egg is set up for visual observation of the developing embryo within. A window will be literally placed on the egg so that the development of the chick can be observed while it incubates. The procedure is not difficult but it does require practice. Try the procedure on an infertile, store-bought egg first. Note any difficult or awkward steps and think of ways to avoid or improve them when working with the fertile egg. Not all of the "fertile" eggs are truly so; some invariably escape fertilization in the hen's oviduct. Before starting, *read* through all the steps.

Drawings of the windowing procedure and of a 72-hour embryo are provided (Figures 8.1, 8.2).

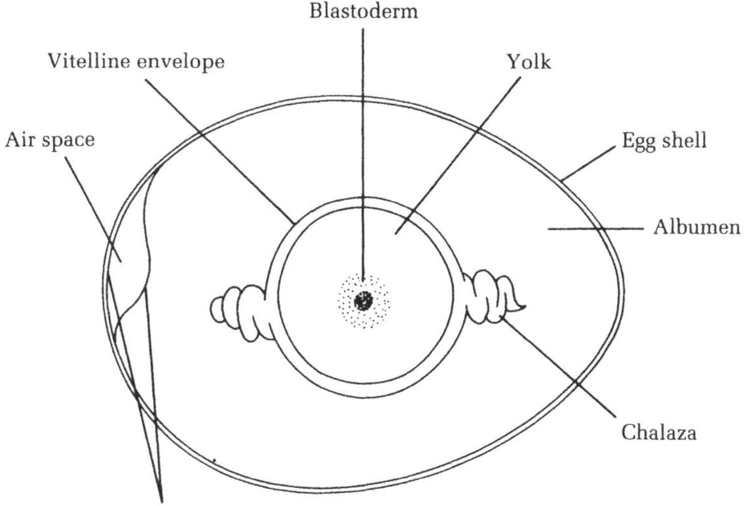

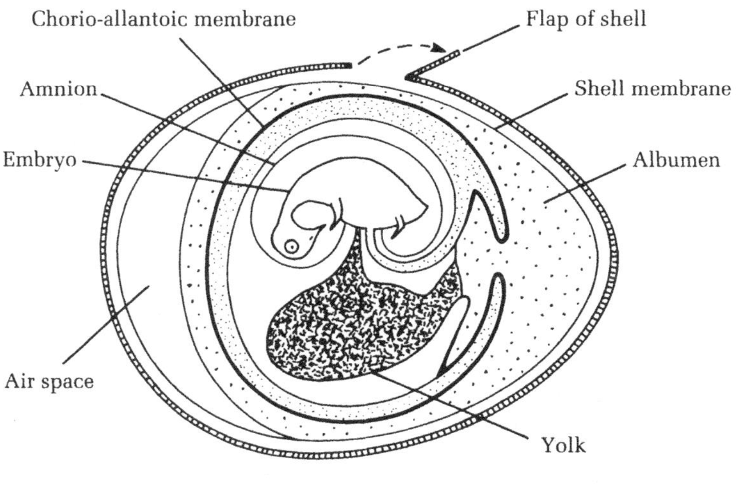

FIGURE 8.1

Schematic drawings of (A) hen's egg; (B) developing chick.

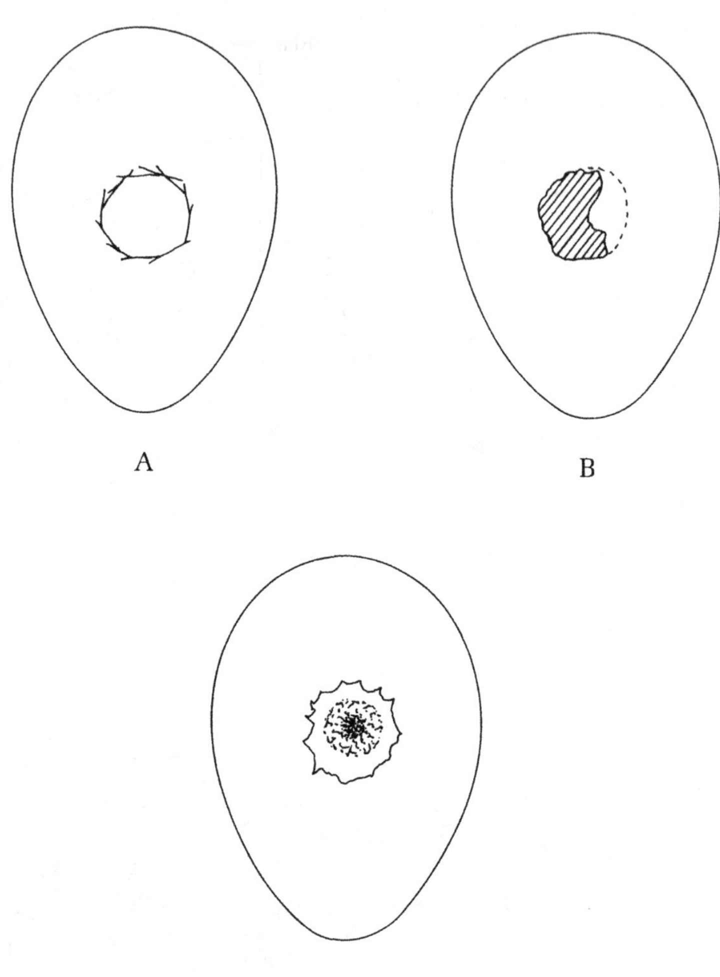

FIGURE 8.2

Steps in windowing an egg. (A) Mark off circle by scoring egg shell with a file. (B) Break off pieces of shell and remove underlying opaque, parchmentlike layer. (C) Expose embryo and use forceps to smooth outline of window by breaking off sharp edges of egg shell.

PROCEDURE

1. Make a viewing stand for the egg by cutting a piece of a styrofoam or cardboard egg carton so that it holds the egg firmly as it lies on its side (broadside, or on its long axis). Try the stand "for size" and for suitability for using with the dissecting microscope. If a prototype stand is available, examine it and make one like it, or make a better one.

2. Obtain an infertile egg. Position it on the stand.

3. Obtain a pencil, a file or miniature saw, coverslips, a paraffin birthday candle, a paraffin applicator (an "unbent" paper clip), an alcohol lamp, and a small flashlight. All of these materials can be shared, except the coverslips.

4. With a pencil (do not try a magic marker — the ink may be toxic), make a circle about 1 cm in diameter on the broad side of the egg. To the left of the circle, write your initials. (Can you guess why the embryo is expected to bob up to the yolk surface so that you can see it?)

5. Wipe the outside of the shell with 70% ethyl alcohol. Wipe all the instruments with alcohol also. Allow the alcohol to *evaporate completely* before you proceed.

6. With a sharpened probe, puncture the shell in the region of the air space (the wide or broad end of the egg; see Figure 8.1). A *small* pinhole is all that is necessary.

7. Turn the egg sideways. Using a small file (or saw), carefully scratch or score short lines along the circle you drew earlier so that pieces of shell can be removed with forceps (Figure 8.2A).

8. With alcohol-sterilized (and dried) forceps, lift off each piece of shell. Peel away the opaque membrane beneath

the shell, if it does not peel off with the shell (Figure 8.2B). Use forceps to smooth outline of window by breaking off sharp edges of egg shell (Figure 8.2C). At the end of this step you should be able to see the blastoderm on top of the yolk when a bright light (use a flashlight) is directed through the window.

9. Check the embryo for developmental stage. *Note it down* in the note pages provided and *make a sketch.*

10. Place a circular coverslip over the window and seal it onto the egg with melted paraffin from a lit birthday candle. Direct the flow of melting paraffin with the un-bent paper clip. As the paraffin solidifies between the rim of the coverslip and the edge of the circular window, it will serve as a gasket to protect the embryo from dehydration and infection. If you are not sure of this step, request a demonstration.

11. Now, do the same with two fertile eggs, marking each with A or B and your initials.

12. Place the labeled, windowed fertile eggs in the incubator. Follow directions from the instructor as to how the eggs should be spaced and positioned in the incubator. Note that the egg is incubated *without* the stand. Keep the stand for subsequent viewings. *Remember the location of your eggs in the incubator.*

13. Examine the embryo in 24 hr and again in 48 hr. Sketch the embryo each time and label the parts that can be viewed through the window. *Important:* Do not open the incubator door or lid unnecessarily because this cools the air within. Move *as quickly as is reasonable* when taking the egg out of the incubator or when placing it inside. When making observations, keep in mind that the egg is cooling down to room temperature. Therefore, shorten the viewing period as much as possible and return the egg promptly to the incubator.

QUESTIONS

1. When you cut out the window, did you expect to find the embryo (or blastoderm) on *that* surface of the yolk, or on the nether surface? Why?

2. Did the embryo develop at the rate predicted by the "chick development" drawing? (Figure 6.1) If not, why not?

3. How far along did your chick develop? The normal time
    to hatching is 21 days.

4. Make suggestions as to how the windowing procedure
    can be improved.

5. What is the function of the air space in the avian egg?
    Would it increase or decrease over time? Why did it
    have to be punctured before the window was made?

References

Biroc, S. L. (1986). *Developmental Biology: A Laboratory Course with Readings,* pp. 46–48. New York: Macmillan Publishing Co.

Rugh, R. (1977). *A Guide to Vertebrate Development,* 7th ed, pp. 213–214. Minneapolis: Burgess Publishing Co.

NOTES TO THE PREPARATOR

1. Fertile eggs can be obtained from most biological supply houses, such as Ward's Natural Science Establishment, Inc. (5100 West Henrietta Road, P.O. Box 92912, Rochester, New York 14692-9012) and Carolina Biological Supply Company (2700 York Road, Burlington, North Carolina 27215). They may also be obtained from commercial farms, for example, Truslow Farms (Route 4, Box 118, Chestertown, Maryland 21620), and from local chicken farms.

2. Incubation is critical and should be timed so that embryos are approximately 75 hr old at windowing. To ensure sufficiently incubated eggs, consider ordering up to 4 eggs per student and then candle the eggs at 72 hr. When viewed through a candling box, the egg should present the image of a cross or a plus sign within a circle. (A candling box is easily prepared by cutting an oval hole approximately two-thirds the diameter of an egg in the top of a box sufficiently large to hold an ordinary desk lamp inside. The desk lamp should not have a shade for maximum brightness. The box itself should be opaque, and the rim of the hole lined with black felt cloth against which the egg can be firmly pressed.)

3. To affix the coverslips to the "windows," use melted wax from hand-held birthday candles. The small size of these candles makes them ideal for this purpose. Coverslips should be previously autoclaved.

4. Windowed eggs will hatch if adequately incubated. Success depends on careful temperature control.

NOTES

NOTES

Ethyl Alcohol, Caffeine, and Chick Development

The incubating chicken egg is an accessible developmental system in which the direct effects of exposure to biologically active substances can be tested. Because the egg is self-contained and the embryo develops autonomously, the effects of different chemical substances applied directly to the developing embryo can be assessed in a straightforward manner and potentially confounding variables, such as maternal effects, can be eliminated. Culture–media effects can likewise be safely ignored because the chicken egg contains all the requisite substances needed by the embryo to complete its development. And because any given chick embryo can be repeatedly observed over the entire course of its development without having to perturb or terminate its progress, time–course studies can be made without requiring different batches of embryos at different stages of development.

Alcohol and caffeine are biologically active substances with known toxic effects at high concentrations. The effects of low concentrations of these substances, however, are perhaps more interesting, particularly during embryonic development, because they can reveal precisely how damage to biological systems is initiated and progresses. In this laboratory exercise, developing chicken embryos will be exposed to low concentrations of ethyl alcohol and caffeine to determine their effects on yolk sac vascularization, blood circulation, and embryonic growth. Figure 9.1 shows a normal chick embryo at 96 hours.

Each lab group should select either ethyl alcohol or caffeine and test their effects at strengths of 0.01%, 0.001%, and 0.0001% on developing embryos beginning at the 96-hr stage. Design the experiment well, keeping in mind the necessity for a rigorously kept control. Discuss how the group will divide responsibilities for bench work and subsequent daily observations for 1 wk. It will be useful to devise a scoring or grading scheme, such as the one that follows, prior to the final evaluation. Daily observations should be recorded in the note pages provided and a report should be submitted 1 wk after the experiment is terminated.

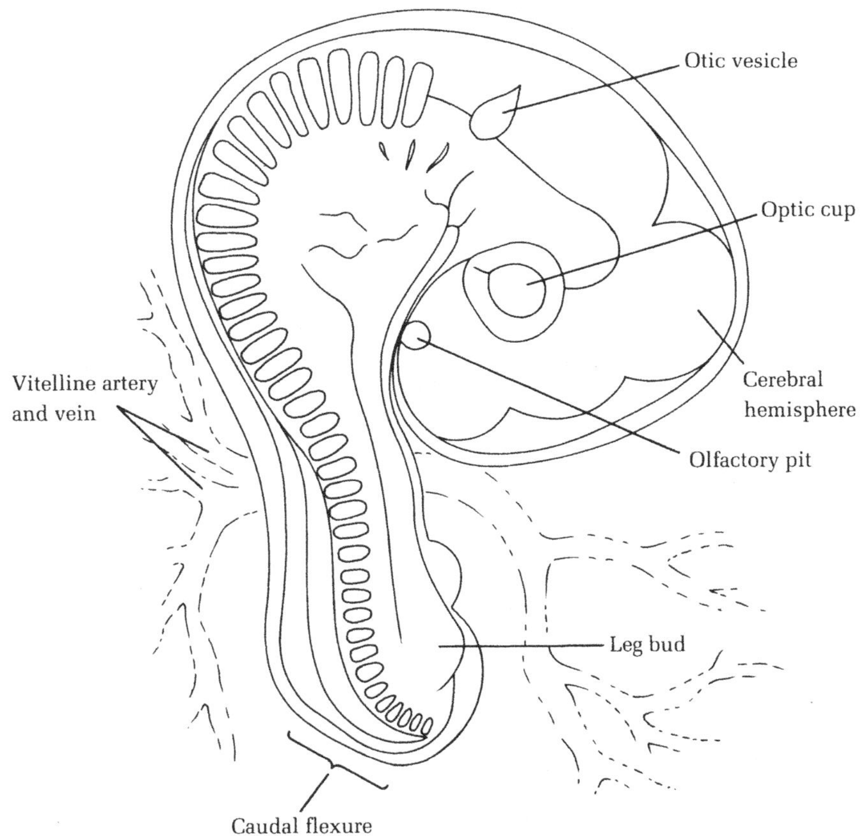

FIGURE 9.1

Chick embryo at ~96 hours.

Example of a Grading Scheme

++   Development strongly enhanced compared with control

+   Development slightly enhanced compared with control

0   No detectable effect (that is, indistinguishable from control)

−   Development slightly impaired compared with control

− −   Development severely impaired compared with control

− − −   Development arrested or embryo killed within 24 hr

PROCEDURE

1. Prepare a "hammock" for incubating 96-hr embryos by modifying a 7-oz clear plastic drinking cup. Pour approximately 30 ml of distilled water into the cup to stabilize it. Cut out a circle of Saran plastic sheeting large enough to form a well 4 cm deep over the mouth of the cup, and still hang about 2 cm over the lip of the cup. Anchor the edges of the plastic sheet by taping it to the outside of the cup. Cap the cup with an airtight lid. Sterilize this assemblage by exposing it to 1–2 hr of ultraviolet radiation. Incubate the sterile "hammock" in a chick incubator.

2. Use only fertile eggs candled at 72 hr to guarantee fertility. Disinfect the egg shell by wiping it first with 70% ethyl alcohol and then with povidone-iodine solution (Betadine). Allow the egg to air-dry with its wide end down. This position causes the embryo to float to the narrow end of the egg, away from the air space, which is located at the wide end.

3. Gently crack the egg at its wide end without puncturing the shell or the parchmentlike shell membrane inside. With a pair of fine forceps, remove pieces of shell and expose the air space all the way to its perimeter. Make the shell edges as smooth as possible to prevent tearing of the vitelline membrane; the egg contents will be poured through this hole later. Figure 9.2 illustrates these steps.

4. The exposed shell membrane can now be manipulated to displace the air space to the narrow end of the egg (that is, just above the embryo; see Figure 9.2). One way to do this is by piercing the exposed shell membrane. Alternatively, tug outward at the shell membrane with a pair of forceps so that its concave surface becomes convex. The latter method provides clear confirmation that displacement of the air space has occurred. A small

amount of albumen may be expelled, but the egg contents should remain in the shell because the displaced air space is still largely a vacuum.

5. Take your "hammock" from the incubator and remove the lid. Hold the egg over the "hammock" with one hand so that the hole faces down into the Saran well.

6. To release the egg contents into the well, crack the narrow end of the egg to allow air to flood the recently created air space. The egg contents should slide out through the exposed shell membrane at the wide end. Should this not immediately happen, rotate the egg along its long axis to ease its contents out. The explanted embryo should be visible on top of the yolk. Replace the cap and incubate the "hammock" while the test solution is prepared.

7. The average chicken egg has a volume of approximately 50 ml. Therefore, the following schedule should yield the desired concentrations if 0.5 ml of the mixture of stock plus Tyrode's solution is used per explanted embryo.

Egg #	Stock caffeine or ethyl alchol (1%)	Tyrode's solution	Final dilution in egg
1	0.5 ml	0	0.01%
2	0.5 ml	4.5 ml	0.001%
3	0.05 ml	4.95 ml	0.0001%
4	0	0.5 ml	0

8. Using a pipette with sterile tips, or a sterile hypodermic needle mounted onto a sterile syringe, add the test solution to the contents of the "hammock." Incubate the explanted embryo for 5 to 7 days.

9. Note your observations in the note pages provided. Discuss your observations with your group and be ready to present a group report in class. Your written report should state the hypothesis being tested, how the testing went, and your conclusions.

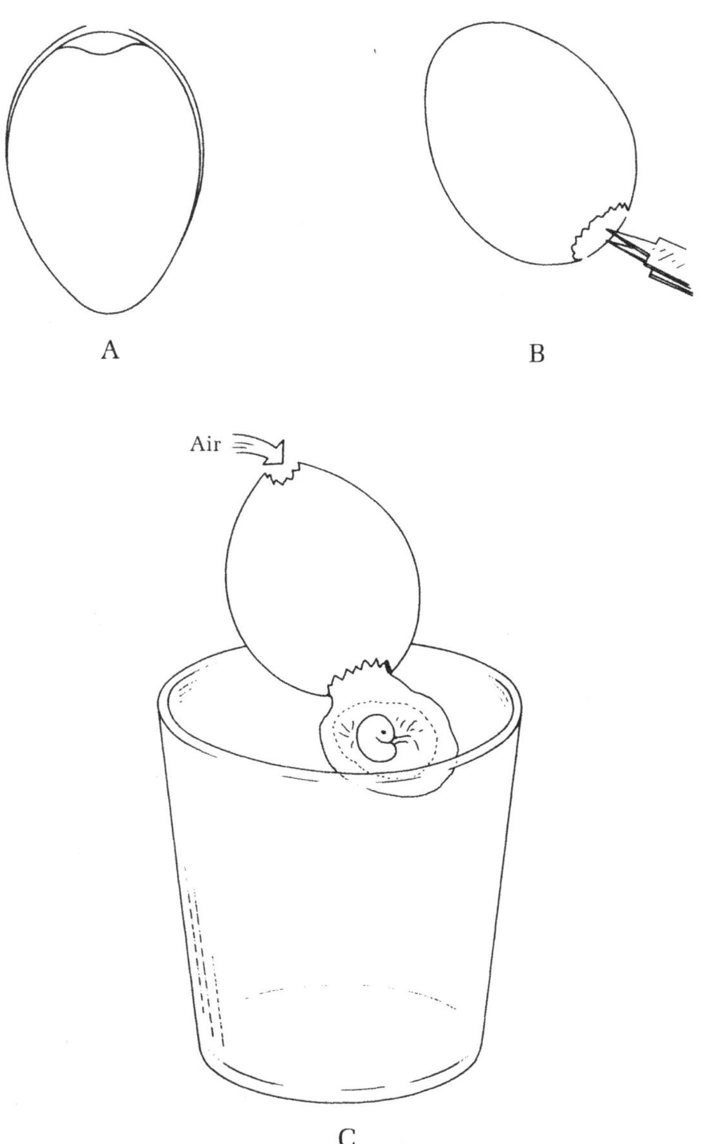

FIGURE 9.2

Explanting contents of chicken egg by first displacing air space from wide to narrow end of egg. (A) Expose egg membrane that covers air space located at wide end of egg. (B) Rupture membrane to displace air space to narrow end of egg. (C) Crack the narrow end of the egg to admit enough air to expel egg contents through window made earlier at wide end of egg.

QUESTIONS

1. What were the observable effects of ethyl alcohol? Caffeine? Are the effects of ethyl alcohol dosage-dependent? What about caffeine?

2. What appropriate controls did you use? How would you adjust for the effects of the explantation procedure alone?

3. What precautions would you take in interpreting your data and extending your conclusions to other experimental systems, such as those of laboratory mice? How about humans?

References

Auerbach, R., Kubai, L., Knighton, D., and Folkman, J. (1974). A simple procedure for the long-term cultivation of chicken embryos. *Dev. Biol.* **41**, 391–394.

Dugan, J. D., Jr., Lawton, M. T., Glaser, B., and Brem, H. (1991). A new technique for explantation and in vitro cultivation of chicken embryos. *Anat. Rec.* **229**, 125–128.

Dunn, B. E., Fitzharris, T. P., and Barnett, B. D. (1981). Effects of varying chamber construction and embryo pre-incubation age on survival and growth of chick embryos in shell-less culture. *Anat. Rec.* **199**, 33–43.

Hamburger, V., and Hamilton, H. L. (1951). A series of normal stages in the development of the chick embryo. *J. Morphol.* **88**, 49–92.

Lehman, H. E. (1977). *Chordate Development,* pp. 252–255. Winston-Salem, North Carolina: Hunter Publishing Co.

NOTES TO THE PREPARATOR

1. Fertile eggs can be obtained from most biological supply houses, such as Ward's Natural Science Establishment, Inc. (5100 West Henrietta Road, P.O. Box 92912, Rochester, New York 14692-9012) and Carolina Biological Supply Company (2700 York Road, Burlington, North Carolina 27215). They may also be obtained from commercial farms, for example, Truslow Farms (Route 4, Box 118, Chestertown, Maryland 21620), and from local chicken farms.

2. Incubation is critical and should be timed so that embryos are 96 hr old for explantation. To ensure sufficiently incubated eggs, consider candling the eggs at 72 hr. The procedure for candling, as well as making a candling box, is given in the *Notes to the Preparator* in Exercise 8.

3. Most plastic cups and lids available in retail stores should suffice, as long as the lids fit snugly. After "hammocks" have been prepared, they may be conveniently UV-irradiated in a tissue-culture (sterile) hood.

4. Caffeine is available from most chemical supply companies, such as Sigma Chemical Company (P.O. Box 14508, St. Louis, Missouri 63178-9916). Povidone-iodine (Betadine) solution is available from Purdue Frederick Company (100 Constitution Avenue, Norwalk, Connecticut 06850-3590).

5. Tyrode's solution may be purchased from GIBCO Laboratories (3175 Staley Road, Grand Island, New York 14072), or prepared as follows:

8.0 g NaCl

0.2 g KCl

0.2 g $CaCl_2 \cdot 6 H_2O$

0.1 g $MgCl_2 \cdot 2 H_2O$

0.05 g NaH_2PO_4

1.0 g $NaHCO_3$

1.0 g glucose

H_2O to make 1 liter

NOTES

NOTES

NOTES

Totipotency of the Unincubated Chick Blastoderm

Totipotency refers to the ability of an embryonic cell to develop into a normal embryo in the absence of other embryonic cells normally associated with it. In many animals, only the zygote is totipotent. However, developmental biologists have known for some time that certain embryos can be manipulated so that their constituent blastomeres, when properly isolated, exhibit developmental potency no less than that of the zygote. Such embryos are said to undergo *regulative development*; that is, developmental potency is retained by each blastomere even after the zygote has cleaved.

The blastoderm is a disc of cells that lies on the surface of the yolky amniote egg. In the fertile chicken egg, the blastoderm typically has attained the blastula stage at the time of oviposition. However, development is temporarily arrested until the egg is incubated at 38.5°C following oviposition. Because the embryonic body axes of the chick are determined after the blastula stage, the unincubated chick blastoderm provides an opportunity to test the totipotency of chick blastomeres at the blastula stage.

In this exercise, the blastoderm of an explanted chicken egg will be bisected and incubated for 1 or 2 days to determine whether each half-blastoderm is capable of producing an embryonic axis, here defined as a notochord, somites, and neural tube. Further incubation may be possible if the manipulated embryos survive bacterial infection. If at all possible, incubation just beyond 2 days should be attempted because of the opportunity to observe the formation of somites, which in the chick proceeds at the rate of one pair per hour. In addition, fusion of the neural folds proceeds at a similarly dramatic rate and is well worth the technical effort demanded by this exercise.

Each lab group should discuss how to proceed with this exercise. It will be useful to practice on infertile, store-bought eggs before doing the procedure on fertile chicken eggs. Special care should be taken at all times to keep the techniques used in this experiment aseptic. Record all observations in the note pages provided. Submit a full report at the conclusion of the experiment.

PROCEDURE

1. Assemble the following materials: UV-irradiated, plastic, cone-shaped cups; Saran plastic sheet to cover the cups and adhesive tape to hold the plastic sheet in place; and cup holders.

2. Make a microsurgical tool by immobilizing a bent No. 000 stainless steel insect pin within the taper of a Pasteur pipette into which melted paraffin has been drawn. If insect pins are unavailable, make your tool by heating the taper of a Pasteur pipette over an alcohol flame and pulling it to a fine filament approximately 10 mm long. Bend the distal 3 or 4 mm of this filament 90° by warming it gently over a small flame.

3. Practice first on an infertile, store-bought egg. Open the egg into a cone-shaped plastic cup. Allow the yolk to orient spontaneously with the blastoderm on the upper surface.

4. With your microsurgical tool, make a small puncture in the yolk on the side of the blastoderm (Figure 10.1A). Insert the needle or pin point under the blastoderm until the point presses against (but does not perforate) the vitelline envelope on the other side of the blastoderm (Figure 10.1B).

5. Carefully retract the needle or pin so that its point cuts through the blastoderm and bisects it (Figure 10.1C). *Do not puncture or slit the vitelline envelope while doing this.* Carefully withdraw the needle or pin from the yolk.

6. Cover the plastic cup with Saran plastic sheet; incubate it for 24 hr. Observe the embryo at the end of this laboratory period, then again at 48 hr. Depending on how aseptic the procedure was, the embryo should survive bacterial infection for about 48 hr. During observation, look for notochords, somites, and neural folds or neural tubes. Carefully record the observations in the note pages provided.

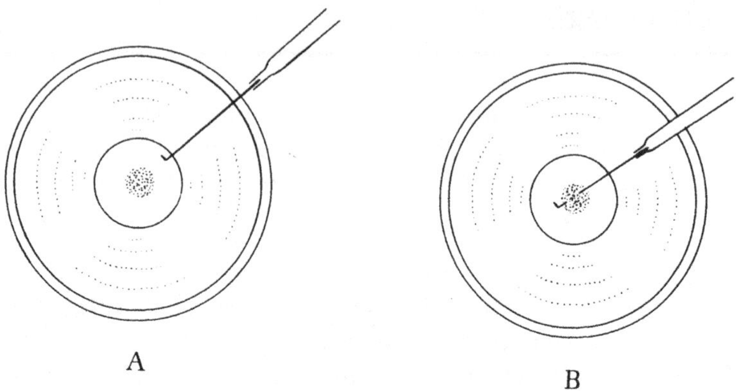

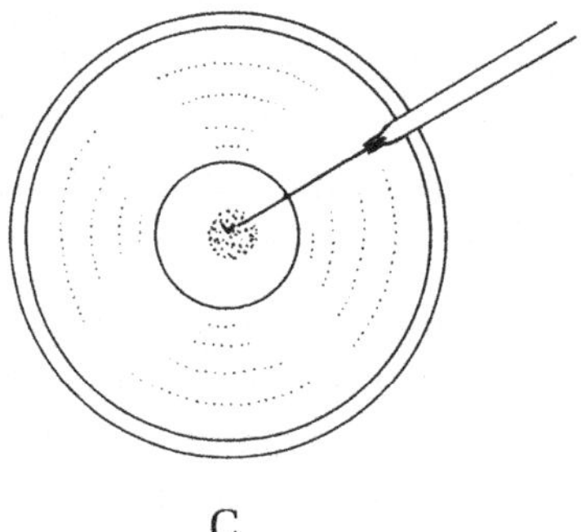

FIGURE 10.1

Bisecting unincubated chick blastoderm with bent insect pin mounted in a Pasteur pipette with melted paraffin. Yolk is shown immobilized in a cone-shaped plastic cup. (A) Insert pin tip through puncture made in the yolk to one side of the blastoderm. (B)Advance pin tip until it touches the vitelline envelope on the other side of the blastoderm. (C) Retract the bent insect pin so its tip slices through the blastoderm without breaching the overlying vitelline envelope.

QUESTIONS

1. Explain how two axes would form from a bisected blastoderm. What does this reveal about the totipotency of cells in the chick blastula?

2. How would you go about determining the maximum number of bisections that a chick blastoderm can sustain and still produce body axes? In other amniotes in which the blastoderm is not flattened (for example, in eutherian mammals), what are the limitations of the bisecting technique used in this exercise?

3. What are the implications of blastoderm bisection for the formation of identical twin embryos?

REFERENCES

Gilbert, S. F. (1991). *Developmental Biology,* 3rd ed, pp. 138–146. Sunderland, Massachusetts: Sinauer Associates.

Lehman, H. E. (1977). *Chordate Development,* pp. 150–151, 189–190. Winston-Salem, North Carolina: Hunter Publishing Co.

Mathews, W. W. (1986). *Atlas of Descriptive Embryology,* 4th ed, pp. 112–121. New York: Macmillan Publishing Co.

Newby, W. W. (1960). *A Guide to the Study of Development,* pp. 18–34. Philadelphia: W. B. Saunders Co.

Slack, J. M. W. (1983). *From Egg to Embryo,* pp. 152–161. Cambridge, England: Cambridge University Press.

NOTES TO THE PREPARATOR

1. Fertile eggs can be obtained from most biological supply companies, such as Ward's Natural Science Establishment, Inc. (5100 West Henrietta Road, P.O. Box 92912, Rochester, New York 14692-9012) and Carolina Biological Supply Company (2700 York Road, Burlington, North Carolina 27215). They may also be obtained from commercial farms, for example, Truslow Farms (Route 4, Box 118, Chestertown, Maryland 21620), and from local chicken farms.

2. Stainless steel insect pins are available from Ward's Natural Science Establishment, Inc. (address given in Note 1).

NOTES

NOTES

NOTES

Cardia Bifida in the Chick

Embryos can be described as either mosaic or regulative. *Mosaic embryos* consist of self-differentiating parts, and as such, they resemble composite entities made of modular units. In contrast, *regulative embryos* consist of interactive, developmentally labile parts that collectively give rise to the embryo. When these parts are artificially disengaged from one another, each part is able to perceive the absence of its peers, redirect its developmental program, compensate for missing parts, and proceed to produce an embryo in the normal manner. In a mosaic pattern of development, the parts differentiate early, remain differentiated even when experimentally disaggregated, and are thus unable to compensate for missing parts. The important distinction between mosaic and regulative embryos, therefore, lies in the cellular and tissue *interactions* that play a major role in the development of the regulative embryo.

Regulative and mosaic development may seem to be mutually exclusive. In reality, they represent the two ends of a continuum. Certain embryos develop in a mosaic manner early on or for most of their developmental period then become regulative; others operate in a regulative mode for a considerable time until eventually they become mosaics of differentiated parts. The transition from regulative to mosaic development occurs slowly in the chick embryo. In the preceding exercise, the totipotency of the unincubated chick blastoderm was examined. This exercise looks at the developing chick heart and examines whether or not its two components, the left and right myocardia, are self-differentiating parts (Figure 11.1).

"Bifid" means forked or split into two component halves. Cardia bifida in the chick arises when the left and right heart primordia fuse unsuccessfully on the ventral median plane. Under normal conditions, the heart forms from two sets of tissues initially located anterior to and to the left and right of the anterior intestinal portal. The objective of this lab exercise is to use microsurgery to prevent the fusion of the primordia to determine whether or not they differentiate individually after 24 hr of incubation.

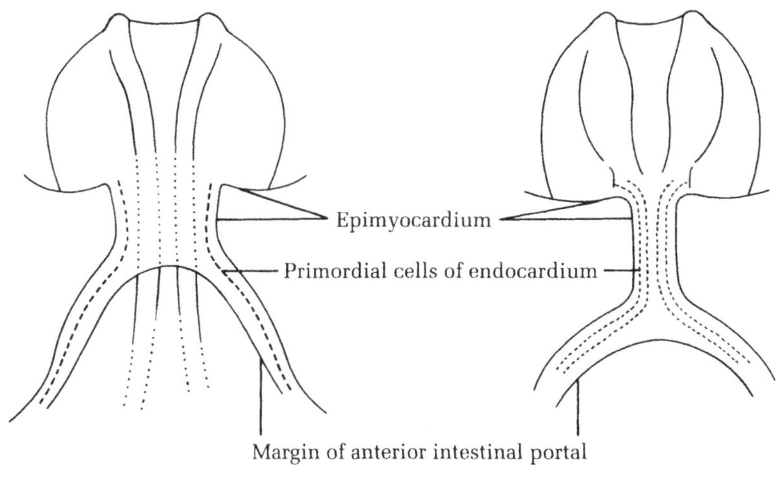

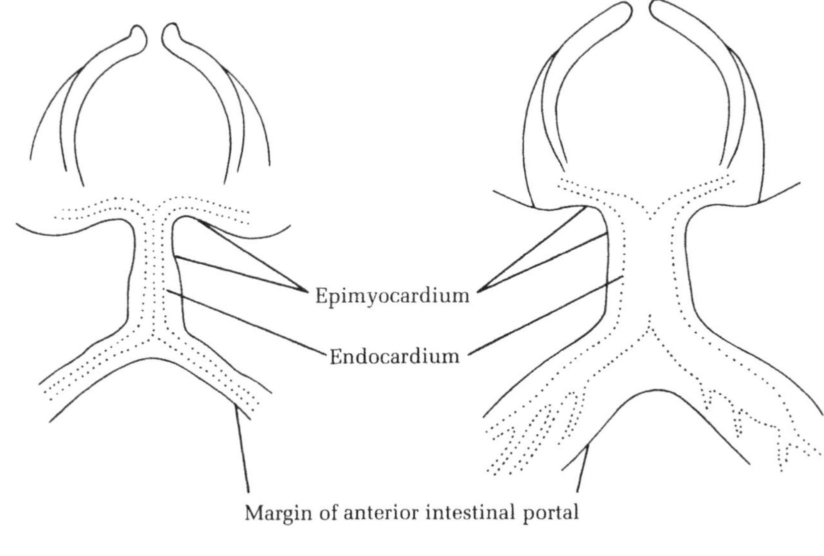

FIGURE 11.1

Diagram of chick embryo heart at 25, 27, 28, and 29 hr of incubation.

Some manual dexterity and practice are required in this lab exercise. Infertile, store-bought eggs will be available for practice in handling the egg yolk and transferring it to a culture dish in which the microsurgery will be performed. Each student will make a microsurgical tool—a heated and pulled glass rod with sufficient rigidity to slice through chick mesodermal tissue. Figure 11.2 illustrates the procedure for this exercise.

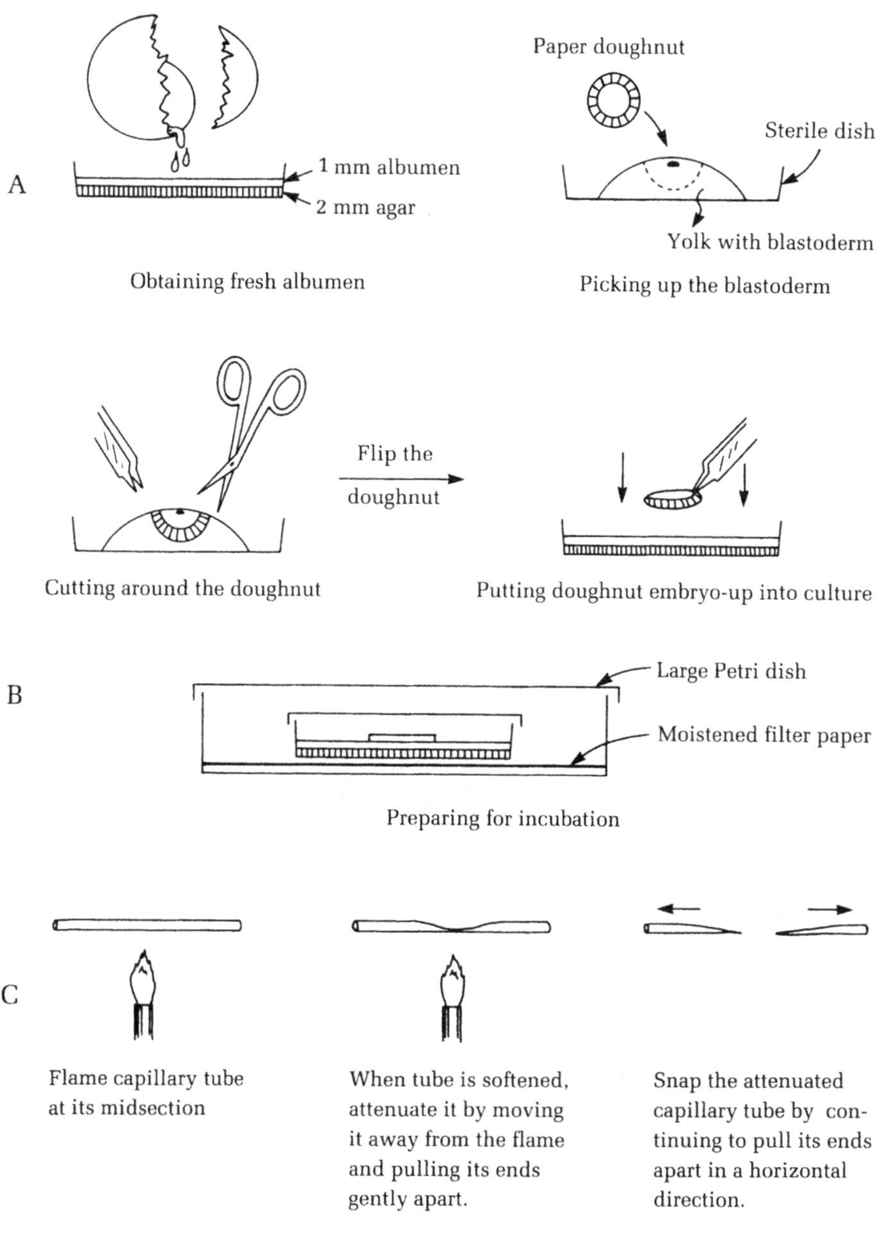

FIGURE 11.2

Helpful tips for (A) explanting 24-hr chick blastoderm, (B) incubating explanted blastoderm, and (C) making microsurgical tool from flame-heated capillary tube.

PROCEDURE

1. Obtain an infertile egg (for practice), two empty (not necessarily sterile) Petri dishes, two (not necessarily sterile) filter paper "doughnuts," a pair of fine forceps, and a pair of scissors (small ones are best). The eggs have been incubated so that the consistency of the contents closely matches that of fertile, incubated eggs.

2. Crack the egg along its "equator," directing the albumen to fill a Petri dish halfway. *Slowly* transfer the yolk between the two half-shells to remove as much albumen as possible. Avoid tearing the vitelline envelope on the sharp edges of the half-shells (Figure 11.2A).

3. Transfer the clean, intact yolk to the second Petri dish. Gently pick or scrape off any lumps of albumen on its surface until it appears almost dry.

4. With forceps, gently position a filter paper doughnut on top of the yolk so that the blastoderm is framed by the doughnut hole. Gently pat the doughnut in place when it fully encircles the blastoderm.

5. While holding the doughnut with the tips of a blunt pair of forceps, cut the vitelline envelope with a pair of scissors and slowly work your way around the rim of the doughnut, carefully checking to see that the vitelline envelope beneath the doughnut adheres to it (Figure 11.2A).

6. Gradually lift the doughnut (at this point framing the blastoderm) and turn it over so that the surface originally facing the yolk now faces the other way. Transfer the doughnut, in this position, to the Petri dish containing albumen (Figure 11.2A).

7. View the explanted blastoderm under the dissecting microscope to get a feel for its size and for the microsurgery to be performed.

8. Now, the real exercise. Each student will assemble the following: a fertile egg, a sterilized Petri dish (top and bottom), a sterile Petri dish with solidified agar albumen, a Petri dish with solidified 4% agar for practice surgery, dissecting scissors and forceps (both swabbed with 70% ethyl alcohol and air-dried on a *clean* paper towel), several melting-point capillary tubes, and a large sterile Petri dish with a filter paper on the floor or bottom of dish (Figure 11.2B). You will share an alcohol lamp, ready for lighting, and a package of sterilized filter paper doughnuts.

9. Break the egg and direct some of the albumen to form a layer about 1 mm thick over the solidified agar albumen. Discard the rest of the albumen. Be careful. Incubated eggs are tricky to handle because the yolk is fatty and becomes runny when heated.

10. Place the egg yolk in a sterile, empty Petri dish. Carefully scrape away all debris until its top looks almost dry.

11. With a sterile pair of forceps, pick up a paper doughnut and proceed to use it, as before, while cutting away the blastoderm.

12. Gently pick up the framed blastoderm, so that any yolk or debris adheres to its yolk-facing side. Invert the blastoderm onto the 1-mm layer of albumen poured over the agar albumen. The blastoderm surface that used to face the yolk should now be facing up into the air.

13. Cover the preparation and write your initials on the cover. Place the covered dish in the large Petri dish with a moistened filter paper on the bottom. Now, place the double dish in a 38.5°C incubator. *Note the time in the note pages provided.* Leave the embryos in for 30 min.

14. While waiting, make the surgical tools. Hold a capillary tube at both ends and heat it near its middle over an alcohol lamp flame until the glass softens (Figure 11.2C.). Immediately withdraw the heated tube from the

flame and pull its ends apart horizontally so that the heated middle portion is stretched until it is about 10 mm long. Briskly snap the resulting thin glass bridge by continuing to pull the ends apart. Each half of the snapped capillary tube should have a very fine but rigid taper about 5 mm long.

15. Practice making the cuts in the agar of the practice surgery dish. You will do the same with the incubating embryo.

16. Set up the dissecting microscope. Remove the embryo from the incubator when its time is up, and sketch it *quickly* in your notebook. (If this takes too long, the embryo will dehydrate and cool down to room temperature.)

17. With the surgical tool, make a longitudinal cut in the *middle of the anterior intestinal portal, through the floor of the foregut.* Pass the glass taper through the foregut two or three times.

18. Cover the inner and outer dishes (both should have your initials) and return them to the incubator. *Record the time.*

19. View the embryo sometime tomorrow (as close as possible to 24 hr). Sketch the embryo and record the time. Do this for every observation of the embryo, returning the embryo to the incubator after observing it. If the microsurgery was successful, there should be two beating hearts in the embryo. Ask the instructor to photograph it.

QUESTIONS

1. Determine how many embryos exhibited cardia bifida among the embryos for which surgery was considered successful. From the successful attempts, reconstruct how the myocardia behave during cardiogenesis.

2. Are the myocardia self-differentiating? How do the results support or weaken this hypothesis?

3. In your report, include a discussion of the transition between mosaic and regulative development in the chick embryo based on the results of this and the previous experiment.

References

Biroc, S. L. (1986). *Developmental Biology: A Laboratory Course with Readings*, pp. 34–45. New York: Macmillan Publishing Co.

Gilbert, S. F. (1991). *Developmental Biology*, 3rd ed, pp. 219–222. Sunderland, Massachusetts: Sinauer Associates.

Mathews, W. W. (1986). *Atlas of Descriptive Embryology*, 4th ed, pp. 125–130. New York: Macmillan Publishing Co.

NOTES TO THE PREPARATOR

1. Fertile eggs may be obtained from biological supply companies such as Ward's Natural Science Establishment, Inc. (5100 West Henrietta Road, P.O. Box 92912, Rochester, New York 14692-9012) and Carolina Biological Supply Company (2700 York Road, Burlington, North Carolina 27215). They may also be obtained from commercial farms, for example, Truslow Farms (Route 4, Box 118, Chestertown, Maryland 21620) and from local chicken farms.

2. Eggs must be incubated for 24 hr. Because the age of the embryo is critical in this experiment, incubate enough eggs to provide students with at least two embryos of the correct age. It will not be possible to candle the eggs because the embryos will be too small to detect.

3. Paper "doughnuts" are made by trimming 1-in diameter filter paper rounds so that only a gasketlike ring of paper remains. The rim of this gasket, if about 3 mm thick, should form a frame around the blastoderm and adhere to the vitelline envelope. This should make the explantation procedure easier and will also prevent the blastoderm from collapsing.

4. Sterile agar albumen is made by warming 5 egg whites in a water bath and slowly pouring it into 100 ml of autoclaved 2% GHR-agar, which is made as follows:

   6 g agar (powder)

   6 g glucose

   150 ml Howard-Ringer's solution

Howard-Ringer's solution is made by dissolving the following in 200 ml double distilled water:

1.44 g NaCl

0.046 g $CaCl_2 \cdot H_2O$

0.075 g KCl

When the agar–albumen solution is homogenized, pour it into sterile plastic Petri dishes to a depth of no more than 2 mm. When cooled down to room temperature, use within 3 hr of preparation.

NOTES

NOTES

Tadpoles and Thyroxine

Amphibian metamorphosis is a dramatic transformation of an herbivorous, aquatic tadpole into a carnivorous, terrestrial adult. This complex process requires a precise orchestration of individual events in the developing tadpole, influenced by triiodothyronine, a secretion of the thyroid gland. The triiodothyronine molecule is composed of two iodinated tyrosine residues. It is therefore a dipeptide, but its small size allows it to traverse the cell membrane and perform as a steroid hormone in modifying gene activity. The effects of triiodothyronine on urodeles (salamanders) are more subtle than they are on anurans (frogs and toads). In the former, the metamorphic changes effected by this hormone include tail resorption, gill degeneration, and skin alterations. Anuran metamorphic changes are considerably more striking, including physiological and structural alterations in the locomotory, respiratory, circulatory, digestive, nervous, and excretory systems. Metamorphosis, in this case, must be a carefully coordinated process.

The coordination of metamorphic events depends on what is called the *threshold concept.* As the concentration of the thyroid hormone rises, different events occur. For example, in water containing dilute amounts of hormone, tadpoles may show only a shortening of the intestines and accelerated hindlimb growth. At higher concentrations of hormone, however, tail regression may occur before hindlimb formation. These observations suggest that metamorphic events occur as a result of organ-specific responses; that is, the response to triiodothyronine is an intrinsic property of the organ itself and does not depend on other factors (such as other organs).

This laboratory exercise tests the hypothesis that exposure to various concentrations of triiodothyronine will elicit different metamorphic changes in the tadpoles of the anuran, *Rana catesbeiana.* Several groups of tadpoles will be constantly exposed, for the duration of the experiment, to different levels of thyroid hormone. The objective of the ex-

periment is to identify the threshold levels for the metamorphic changes that are observable. These changes should be
noticeable in such externally visible structures as the tail
and limbs. Tadpoles will be dissected at the end of the experiment to determine any changes in their gills and intestines.

An important component of this lab exercise is the cooperation within and among the different student groups who
will be performing specific segments of the experiment. Coordination of feeding, water-changing, and tadpole-measur-
ing chores is necessary for the success of the experiment,
which will run from 4–6 wk. Measure tadpoles once a week,
just before or just after lab class. Feeding and water changing
must be done *daily*; this includes over spring break and
weekends. A great deal of common courtesy and trust is required for this lab because water pails have to be filled and
labeled accurately, thyroxine solutions have to be stored
properly, and so on. If not, all the elaborate preparations
could yield uninterpretable results—the ultimate horror for
any experimentalist!

All observations should be recorded in the note pages provided. These will include measurements, descriptions of
any visible morphological or behavioral changes in the tadpoles, and any variations in their care and feeding. At the
end of the experiment, the class will pool the data. The primary purpose of the lab reports is to interpret the pooled
data.

PROCEDURE

1. Divide the class into four groups: I, II, III, and IV.

2. Each group should obtain the following: 2 aquaria with 5 liters of dechlorinated tap water in each, 5 sheets of millimeter graph paper, 2 finger bowls per group, and 1 small aquarium net to use for tadpoles.

3. Mark the 5-liter water level on the outside of each aquarium with a permanent marker so that the amount of water will not have to be measured each time the the contents are changed. Label the aquariums as follows:

   Group I = Control

   Group II = 10^{-5} M thyroxine

   Group III = 10^{-6} M thyroxine

   Group IV = 10^{-7} M thyroxine

4. Use an aquarium net to transfer 5 tadpoles to each aquarium.

5. Each student should take out 1 tadpole from either aquarium, and transfer it into a finger bowl with refrigerated or ice water. The cold water will anesthetize the tadpole, slowing it down so that observations can be made. (*Important:* Gradually add the ice cubes. If the water is cooled too rapidly, the shock could kill the tadpoles.) Slide a sheet of millimeter graph paper under the finger bowl to serve as a measuring grid. An alternative way to measure the tadpoles is to use a short ruler calibrated in millimeters and hold it near the tadpole while it is in the finger bowl.

6. Using a net, return the tadpole to its aquarium. Add a small amount of spinach leaf and rabbit chow for the tadpoles to eat. Watch the tadpoles feed and from this, estimate how much they can eat in a 24-hr period. *Put in just as much spinach as the tadpoles will eat in 24 hr; otherwise the spinach rots and fouls the water.*

7. Keep the thyroxine stock solution in foil-wrapped or opaque glass bottles in the refrigerator. Shake each bottle well, but carefully, before taking out the correct amount of stock solution for each aquarium, as indicated. *Remember to return the thyroxine bottle to the refrigerator when you are done.*

   Group I = control (no thyroxine)

   Group II = 5 ml thyroxine stock solution *per aquarium*

   Group III = 0.5 ml thyroxine stock solution *per aquarium*

   Group IV = 0.05 ml thyroxine stock solution *per aquarium*

8. Set the aquaria aside and adjust the aerators. Observations must be made next week on the same day. Record observations in your notebook. The experiment runs 4 wk, or longer if necessary.

9. The tadpoles have to be *fed daily* and have their *water changed every day.* Replenish their food every day, providing just enough spinach every day to last 24 hr. Make a schedule for each group so that all members participate equally. Remember to add thyroxine to the water for Groups II–IV.

10. It is the responsibility of each group to record the following observations for each group of tadpoles: (a) head width (mm); (b) tail length (mm); (c) total body length (mm); (d) appearance of limbs, tail, and so on; (e) activity level; and (f) deaths, if any. At the end of the experiment, the class will collate the data and discuss the results.

11. In the event that a tadpole dies, the following should be noted: when and how it died (your best guess). The tadpole should be fully described in your report, and any metamorphic changes (described in the introduction to this lab exercise) should be recorded. This means dissecting the tadpoles as soon as possible. Tadpoles may be preserved in 70% ethanol if dissection cannot be performed immediately.

12. At the conclusion of the experiment, surviving animals will be sacrificed and dissected in order to compare metamorphic changes occurring in visceral organs as a result of tadpole exposure to different concentrations of thyroxine.

QUESTIONS

1. How would you describe the emergence and growth of the limbs? The changes in the tadpoles' tails? Did limb changes precede tail changes?

2. Make a graph to represent each set of measurements. Devise a way of reporting the limb emergence data. What trends, if any, are noticeable in your graphs?

3. Do the data support the threshold concept? Explain your answer. In your report, include a discussion of the threshold concept as it applies to this experiment.

References

Biroc, S. L. (1986). *Developmental Biology: A Laboratory Course with Readings,* pp. 197–198. New York: Macmillan Publishing Co.

Gilbert, S. F. (1991). *Developmental Biology,* 3rd ed., pp. 686–699. Sunderland, Massachusetts: Sinauer Associates.

Kollross, J. J. (1961). Mechanisms of amphibian metamorphosis: Hormones. *Amer. Zool.* **1,** 107–114.

NOTES TO THE PREPARATOR

1. Hind-leg-emergent bullfrog tadpoles are available from most biological supply companies, such as Ward's Natural Science Establishment, Inc. (5100 West Henrietta Road, P.O. Box 92912, Rochester, New York 14692-9012) and Carolina Biological Supply Company (2700 York Road, Burlington, North Carolina 27215).

2. To dechlorinate tap water and to ensure that it will be at room temperature when used, allow it to stand for about 12 hr in an open container.

3. L-thyroxine is available in powder form from Sigma Chemical Company (P.O. Box 14508, St. Louis, Missouri 63178-9916). It should first be dissolved in a small volume of 95% ethyl alcohol, then slowly mixed into water to a final stock concentration of 0.1 M. If thyroxine precipitates out of the solution, add by drops, 1 M NaOH to increase the pH and improve solubility. The stock solution should be stored in a dark or foil-wrapped glass bottle and refrigerated. It should be shaken thoroughly before using.

4. Because this lab exercise runs over many weeks and the aquarium water has to be changed daily, students should have off-hours access to the lab and to such materials as tadpole food, thyroxine stock solution, and dechlorinated water.

5. Class data should be pooled and analyzed at the end of the experiment.

NOTES

NOTES

NOTES

NOTES

NOTES

Blastomere Totipotency in the Two-Cell Mouse Embryo

Development can be viewed as the stepwise loss of totipotency at the cellular level. Embryonic development therefore represents a gradual decline in the number of developmental options available to a cell or tissue. In embryos that undergo mosaic development, this decline occurs at a very early stage. Consequently, blastomeres proceed with fair autonomy along fixed developmental pathways, seldom deviating, whether or not they are in the presence of other blastomeres. On the other hand, embryos that undergo regulative development consist of blastomeres that remain developmentally labile through the cleavage stages and sometimes beyond. As a result, isolated blastomeres from these embryos will frequently compensate for missing peers and proceed to develop normally.

By definition, the mouse zygote is developmentally totipotent (Figure 13.1). During cleavage, however, the blastomeres in a mouse embryo predictably undergo the gradual decline in totipotency characteristic of regulative development. In theory, therefore, it is possible to determine whether or not a blastomere of a given age remains totipotent. In practice, this can be accomplished by isolating cleavage-stage blastomeres and finding out whether or not they resume normal development.

This exercise will test the hypothesis that the blastomeres in a two-cell mouse embryo are totipotent by dissolving the zona pellucida, disaggregating the blastomeres, and observing whether or not they develop to the blastocyst stage. If these blastomeres are totipotent, each of them should form a normal blastocyst more or less on schedule. Blastocysts formed by isolated blastomeres will be compared with those formed by zona-free, otherwise intact, two-cell embryos. If the hypothesis is correct, blastocysts formed by isolated blastomeres should be indistinguishable from those formed by two-cell embryos.

This exercise requires a fair amount of technical skill. The embryos are no more than 100 microns in diameter and require strict aseptic handling. In addition, these embryos

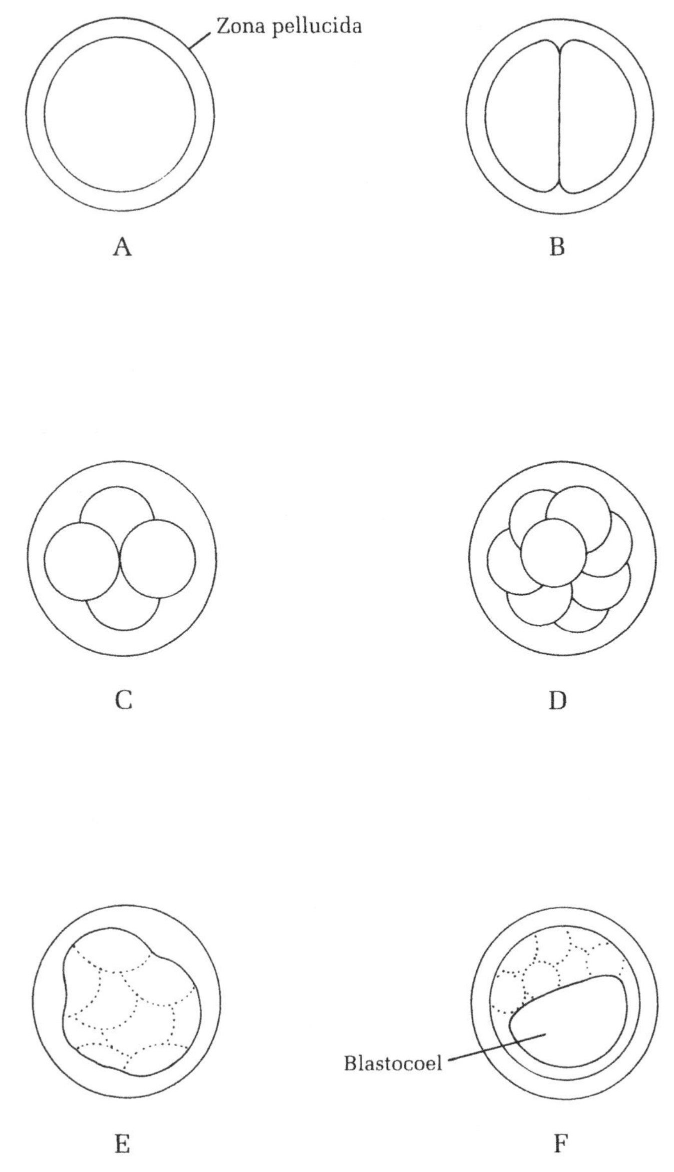

FIGURE 13.1

Mouse embryo preimplantation stages. (A) One-cell embryo. (B) Two-cell embryo. (C) Four-cell embryo. (D) Eight-cell embryo. (E) Morula. (F) Blastocyst.

must stay as much as possible at 37°C, which means working quickly to restore the embryos to this temperature by taking the minimum amount of time for manipulations and promptly returning the embryos to an incubator. Needless to say, practice, organization, and patience are prerequisites for successful completion of the experiment.

Students will work in groups, dividing tasks to maximize efficiency and optimize results. Read the following procedure first, then set up a work space and conduct a dry run or two. Discuss within the group what the appropriate controls should be. Record the results in the note pages provided and be able to present a summary of the group's work and findings to the class. The report should include a discussion of the implications of the results.

PROCEDURE

1. Set up a work station by assembling the following items: a trans-illuminated dissecting microscope, sterile plastic Petri dishes, sterilized Pasteur pipettes, an alcohol lamp, and a micrometer syringe fitted with rubber tubing of sufficient diameter to accept the wide end of a Pasteur pipette. Practice pulling Pasteur pipettes to a fine taper using an alcohol flame. Use the pulled pipettes, modified by the attached micrometer syringe, to transfer embryos from one solution to another.

2. Obtain a small dish of embryos and a dish with a drop of Pronase in it. Transfer the embryos into the Pronase dish, cover the dish, and, while viewing the embryos through the dissecting microscope, observe the dissolution of the zona pellucida. Pronase is a protease which will dissolve the zona pellucida in 5–10 min at the concentration that will be used here.

3. As soon as the zona pellucida is dissolved, rinse the embryos in a preincubated culture medium poured into a sterile Petri dish. Before proceeding, allow the embryos to recover for about 1 hr in an incubator held at 37°C, 100% relative humidity, and 5% CO_2 atmosphere.

4. Transfer the embryos into a dish of preincubated, calcium-free medium overlain with paraffin oil (to prevent evaporation, which would increase the osmolarity of the medium). Incubate the embryos for 10–15 min; meanwhile, prepare flame-polished glass Pasteur pipettes for dissociating the embryos. Prepare the holding dishes to be used for the isolated blastomeres by spacing 30-µL drops of calcium-containing medium in a sterile tissue culture dish. Overlay these drops (without disrupting any of them) with paraffin oil. Preincubate the holding dishes after making sure that there are enough drops (one drop per blastomere).

5. Disaggregate the blastomeres by pipetting them several times through the prepared Pasteur pipettes. Some paraffin oil will cling to the Pasteur pipette, but will not touch the blastomeres if they are kept fully immersed in culture medium *under* paraffin at all times.

6. Remove the blastomeres from the calcium-free medium as soon as possible. The isolated blastomeres are sticky; handle them delicately and transfer them individually into the drops of medium previously prepared in the holding dishes. Paraffin oil adhering to the transfer Pasteur pipette will mix with the oil overlay in the holding dishes but not with the culture medium.

7. Remember to set up controls.

8. Observe the embryos and blastomeres over the next 48 hr. Record observations in the note pages provided. Interpret the results and discuss their implications (totipotency, regulative development, cloning, twinning, etc.) in the report.

QUESTIONS

1. Why was it necessary to transfer the embryos to a calcium-free medium for disaggregation?

2. How do the sizes of blastocysts formed by isolated blastomeres compare with those formed by zona-free, 2-cell embryos? Why is there or why is there not a size difference? Discuss the implications of observed size differences (or lack thereof) in terms of regulative embryonic development.

3. What is a chimera? How would you modify the technique described in this exercise to obtain chimeric embryos?

References

Gilbert, S. F. (1991). *Developmental Biology,* 3rd ed, pp. 89–98. Sunderland, Massachusetts: Sinauer Associates.

Hogan, B., Costantini, F., and Lacy, E. (1986). *Manipulating the Mouse Embryo: A Laboratory Manual,* pp. 106, 111, 249–277. Cold Harbor Spring, New York: Cold Spring Harbor Press.

Pedersen, R. A. (1986). Potency, lineage, and allocation in preimplantation mouse embryos. In *Experimental Approaches to Mammalian Embryonic Development,* J. Rossant and R.A. Pedersen, eds., pp. 3–33. Cambridge University Press.

Pratt, H. P. M. (1987). Isolation, culture and manipulation of preimplantation mouse embryos. In *Mammalian Development: A Practical Approach.* M. Monk, ed., pp. 13–42. Oxford and Washington, D.C.: IRL Press.

Quinn, P., Barros, C., and Whittingham, D. G. (1982). Preservation of hamster oocytes to assay the fertilizing capacity of human spermatozoa. *J. Reprod. Fertil.* **66**, 161–168.

Smith, R., and McLaren, A. (1977). Factors affecting the time of formation of the mouse blastocoele. *J. Embryol. Exp. Morphol.* **41**, 79–92.

Tsunoda, Y., and McLaren, A. (1983). Effect of various procedures on the viability of mouse embryos containing half the number of blastomeres. *J. Reprod. Fertil.* **69**, 315–322.

Ziomek, C. A., and Johnson, M. H. (1980). Cell surface interaction induces polarization of mouse 8-cell blastomeres at compaction. *Cell* **21**, 935–942.

NOTES TO THE PREPARATOR

1. Laboratory mice are available from many commercial suppliers, such as The Jackson Laboratory (Bar Harbor, Maine 04609) and Harlan Sprague Dawley, Inc. (P.O. Box 29176, Indianapolis, Indiana 46229).

2. Female mice to be used as embryo donors should be 6–8 wk old and ideally entrained to a 14:10 light–dark cycle so that their estrus cycles coincide. Each is super-ovulated with two intraperitoneal injections 48 hr apart, first with pregnant mare serum gonadotropin and then with human chorionic gonadotropin at 5 IU per female. Both of these hormones are available from Sigma Chemical Company (P.O. Box 14508, St. Louis, Missouri 63178-9916). Immediately following the second injection, each female is placed in the same cage with a breeding male overnight. The following morning, ideally before 10 A.M., the females should be checked for copulation plugs; those without plugs very likely did not mate and will therefore not yield embryos.

3. To obtain 2-cell embryos, explant the oviducts and flush them from the ampulla end with a stream of pre-warmed flushing medium introduced by a syringe fitted with a trimmed (7 mm) and finely smoothed 33-gauge needle (Hamilton Company, P.O. Box 10030, Reno, Nevada 89520). Embryos are handled using an aspirating device made from a micrometer syringe fitted with rubber tubing of sufficient diameter to accept the wide end of a Pasteur pipette. The Pasteur pipette should first be pulled over a flame to produce a fine taper no more than 200 microns in diameter. Rotating the micrometer knob back and forth will create sufficient suction to aspirate embryos into the pipette taper or adequate force to expel the embryos gently. (Mouth suctioning is possible but not highly recommended.)

4. All containers used for handling embryos and culture media should be sterile. Gamma-irradiated plastic Petri dishes are available from many general laboratory supply companies. A water-jacketed incubator with a carbon dioxide (CO_2) supply to maintain a humid, interior atmosphere of 5% CO_2 should be used for both embryos and media. Ensure that relative humidity within the incubator is 100% by placing a pan of sterile distilled water on the floor of the incubator.

5. Embryos are best viewed using a binocular dissecting microscope with transillumination. Visibility of the zona pellucida is greatly enhanced by transillumination.

6. The ingredients for culture media and Pronase solution, as well as paraffin oil, are available from Sigma Chemical Company (P.O. Box 14508, St. Louis, Missouri 63178-9916).

Culture Medium (Modified Krebs-Ringer Solution; Quinn *et al.* 1982)

94.66 mM NaCl

4.78 mM KCl

1.71 mM $CaCl_2 \cdot 2H_2O$

1.19 mM KH_2PO_4

1.19 mM $MgSO_4 \cdot H_2O$

4.15 mM $NaHCO_3$

20.85 mM HEPES

23.28 mM sodium lactate

0.33 mM sodium pyruvate

5.56 mM glucose

4 g/liter bovine serum albumin

0.06 g/liter Penicillin G (potassium salt)

0.05 g/liter streptomycin sulfate

0.01 g phenol red

Up to 1 liter double distilled water

Dissolve HEPES in 50–100 ml of double distilled water. Adjust pH to 7.4 with 0.2 N NaOH. Dissolve antibiotics in double distilled water; do the same with $CaCl_2$. Dissolve remaining components (except bovine serum albumin and lactate) into a 1-liter volumetric flask and add 500 ml of double distilled water. Add antibiotics, HEPES, and $CaCl_2$ to the volumetric flask. Weigh out the lactate syrup into a 10-ml beaker and add to the flask. Rinse the beaker several times with double distilled water, adding the washings to the flask. Make up the volume to 1 liter. Sprinkle bovine serum albumin on top of the medium and allow to dissolve slowly. *Mix gently, do not shake.* If necessary, readjust the pH to 7.2–7.4 with NaOH. Filter through a 0.22 micron Millipore filter, discarding the first few milliliters. Store at 4°C for up to 1 wk. The osmolarity should be 285–287 milliosmol.

7. Pronase solution consists of 0.5% Pronase in a modified Krebs-Ringer solution.

NOTES

NOTES

NOTES

Alizarin Staining

An area in developmental biology of interest to biomedical and environmental scientists, as well as to pharmacologists and toxicologists, is teratology, the study of morphological abnormalities that arise during embryogenesis. Teratogenic agents can be chemical (antibiotics, sodium nitrite, thalidomide, and alcohol) or physical (x rays and ultraviolet rays). These agents disrupt the timing of developmental events, thereby causing physical malformations.

One of the techniques employed in the assessment of bone and cartilage defects is alizarin–alcian staining. Alizarin red stains osteogenic (bone-producing) tissues red. Alcian blue binds to chondrogenic (cartilage-producing) tissue and causes the tissue to turn blue. Muscle, skin, nerves, and blood vessels obscure any staining of bone and cartilage within, so these soft tissues are rendered transparent by chemical treatment. In this manner, stained bone and cartilage can be examined *in situ*.

In this lab exercise newborn mouse pups will be processed for alizarin–alcian staining of cartilage and bone. This procedure takes several days because the various chemicals require time to penetrate the tissues. Each student will be provided one pup to work with, and will be responsible for all procedures done outside of laboratory periods. In the note pages provided, make and record observations at each step to chart the progress of the staining procedure.

Important: Each student will often be working in the lab alone for the duration of this exercise. Please be considerate of other students and clean up after working. Breakage and spills should be reported when they occur and immediately cleaned up. Because you will be using various chemicals, *make certain* you know what precautions are necessary in their handling and disposal. When unsure, err on the side of caution. Better yet, ask for help or advice.

PROCEDURE

1. Obtain an eviscerated mouse pup preserved in 4% saline. With gloved hands, take out the pup and place it on a Petri dish. Using fine forceps, carefully remove the skin. Dispose of removed tissues in designated containers.

2. Do this step in the hood with the fan on. Transfer the specimen into a vial of 10% formaldehyde in phosphate-buffered saline. (*Note*: **Formaldehyde is a carcinogen. Do not handle with bare hands.**) Leave specimen in tightly covered vial overnight in the hood. This will fix the tissues.

3. Wash the specimen in 6–10 changes of distilled water over a 2-day period. This will remove most of the formaldehyde from the tissues. Remember to discard the washings into a formaldehyde disposal jar. Do not wash formaldehyde down the sink.

4. Transfer the specimen into a vial containing Alcian blue solution so that the animal is completely submerged. In 12–24 hr, all superficial cartilage will turn blue. You must check this every 6–10 hr, so as not to overstain. *DO NOT LEAVE SPECIMENS IN ALCIAN BLUE STAIN LONGER THAN NECESSARY BECAUSE OSSIFIED TISSUES WILL DECALCIFY.*

5. Decant the Alcian blue solution into the sink, while the tap is running steadily. Pour in enough absolute ethanol into the vial so that the specimen is completely submerged. Leave in alcohol for 24 hr.

6. Decant the alcohol into the sink. Replace with a second alcohol bath; leave for another 24 hr. This will dehydrate the specimen, fix the Alcian blue in the cartilage, and help destain noncartilaginous tissue.

7. Decant the second alcohol bath. Pour in each of the following solutions (after decanting each previous one) at 2- to 2.5-hr intervals: 75%, 50%, and 25% ethanol; 2 baths of distilled water.

8. Pour in 0.5% potassium hydroxide (KOH), and allow soft tissues to become transparent (this is called the maceration step). This may take up to 3 days. Check with the instructor to determine when clearing is sufficient.

9. Decant the KOH solution; replace with Alizarin red solution. Leave for 24 hr. When stained, bone will appear red to purple.

10. Decant alizarin red solution. Replace with the following solutions, in order, every 24 hr: 25%, 50%, 70% glycerin in 0.5% KOH. [*Note:* To the 25% glycerine-KOH bath, add one drop of 3% hydrogen peroxide (H_2O_2) in order to bleach any skin or pigmented connective tissue left in or on the specimen.]

11. Finally, transfer the specimen carefully, using forceps, to a vial of pure glycerin to which a few crystals of phenol or thymol have been added to prevent spoilage. **Do not handle crystals with bare hands; use a spatula or forceps.**

QUESTIONS

1. Did any structures in your specimen stain blue in certain areas and red in others? What might be the reason for this?

2. Alizarin staining is a method with varied applications. Suggest two research problems (other than evaluation of teratogenicity) that may be investigated using this technique.

References

Hanken, J., and Wassersug, R. (1981). The visible skeleton. *Funct. Photog.* **16**, 22–26, 44.

Simons, E. V., and Van Horn, J. R. (1971). A new procedure for whole-mount alcian blue staining of the cartilaginous skeleton of chicken embryos, adapted to the clearing procedure in potassium hydroxide. *Acta Morphol. Neerl.-Scand.* **8**, 281–292.

Wassersug, R. J. (1976). A procedure for differential staining of cartilage and bone in whole formalin-fixed vertebrates. *Stain Technol.* **51**, 131–134.

NOTES TO THE PREPARATOR

1. Laboratory mice are available from many commercial suppliers, such as The Jackson Laboratory (Bar Harbor, Maine 04609) and Harlan Sprague Dawley, Inc. (P.O. Box 29176, Indianapolis, Indiana 46229). Keep 6- to 10-week-old females in cages with males, checking for copulation plugs daily before 10 A.M.The day of plug detection is day 1 of pregnancy.

2. Use mouse pups Caesarian-delivered on day 20 of pregnancy. During explantation, remove the placenta and all other extraembryonic tissues. Eviscerate pups, being careful to remove all internal organs. Rinse eviscerated pups in 4% saline and store refrigerated for 18–24 hr before using. This period of refrigeration in saline is required to facilitate skin removal.

3. Alcian blue and Alizarin red (Sigma Chemical Company, P.O. Box 14508, St. Louis, Missouri 63178) solutions are prepared as follows:

    Alcian blue solution

    40 mg Alcian blue

    140 ml absolute ethyl alcohol

    60 ml glacial acetic acid

    Alizarin red solution

    100 ml 0.1% Alizarin red in distilled water

    400 ml 0.5% KOH

4. Phosphate-buffered saline is prepared by mixing 50 ml 0.2 M dibasic sodium phosphate with sufficient 0.2 M monobasic sodium phosphate to obtain a pH of 7.20–7.25.

NOTES

NOTES

NOTES

Drosophila Polytene Chromosomes

The fruit fly, *Drosophila melanogaster* (literally, "dark-bellied dew lover"), is a classical research tool in genetics. Every biologist alive today knows about *Drosophila* genetics and chromosomes. Over the past 20 years, this insect has emerged as an eminent system for studying animal development because of its short life cycle, ease of manipulation, and, most significantly, because of the vast store of genetic information that has accumulated in the nearly 100 years that this fly has been used in laboratories. Because most biological functions in all organisms share common or similar genetic bases, *Drosophila* has also contributed to the growth of neuroscience, biochemistry, molecular biology, cell and developmental biology, evolutionary biology, entomology, and now, genetic engineering.

Possibly the most distinctive characteristic of *Drosophila* (and certain related flies) is the *polytene* (multistranded) chromosome. Each strand is a faithful copy of the original DNA, and is matched, point by point along its entire length, to neighboring copies. Each chromosome represents about 1000 duplexes of DNA, held in perfect register with one another. Certain regions are more tightly packed than others and give rise to the distinctive transverse *bands* of these chromosome formations when stained with DNA-sensitive dyes. For example, the banding pattern is invariant for chromosome *3*; consequently, all such chromosomes of every *Drosophila melanogaster*, anywhere at anytime, will look identical, barring mutations involving large chunks of DNA. Another special feature of polytene chromosomes is the *somatic pairing* that homologous chromosomes undergo. This means that, for example, the two chromosomes *3* are intertwined so that only two, not four, chromosome arms appear to project from the region of the centromere. A final special feature of polytene chromosomes worth mentioning here is the *chromocenter,* or the common intersection of each of the intertwined homologs' centromeres. The chromocenter appears granular under the compound microscope in well-stained chromosome squashes and therefore probably consists of special proteins that hold the polytene chromosomes in a bundle.

Drosophila melanogaster has a diploid chromosome number of 8. However, in cells with polytene chromosomes, there are only four cablelike formations (due to somatic pairing) bundled together at their centromeres (the region called the chromocenter). Chromosomes *1* (which is the *X* chromosome) and *4* (which is the *Y* chromosome, and a very small one at that) are acrocentric; therefore these chromosomes appear to project from the chromocenter, anchored in by the centromeric ends. Chromosomes *2* and *3* are metacentric; thus there are arms called "*2L*," "*2R*," "*3L*," and "*3R*." If one were to count arms, therefore, every nucleus containing polytene chromosomes would appear to have six arms emanating from the chromocenter.

In this lab exercise, the salivary glands from *Drosophila melanogaster* larvae will be excised (not all *Drosophila* cells have polytene chromosomes) and processed to reveal the banding patterns of these giant chromosomes (Figure 15.1). Take note of any morphological variation along the length of the different chromosome *arms* and identify the *puffs,* local regions of strand detachment or unraveling (other picturesque terms that have been used are ballet skirt, Chinese lantern, goose neck, duck's head, onion base, turnip, and shoe buckle), and the *loops.* In the note pages provided, draw the best chromosome squashes in the specimen. Under oil immersion, view the telomeres of each chromosome arm and identify the arms based on their banding patterns. A chart illustrating banding patterns should be available in the laboratory.

PROCEDURE

1. Set up both microscopes. Obtain a glass slide, several coverslips, absorbent paper squares, and a wad of tissue paper.

2. With a partner, set up one bottle each of saline, aceto-orcein, immersion oil, acetic acid, and nail polish. Share an alcohol lamp.

3. Place a drop of saline (approximately 5 mm diameter) on a glass slide. Obtain a last-instar larva (maggot) from the culture jar. The ones crawling up the wall are best. Do not get pupae, and do not release adult flies from the bottles! (If you have to etherize the flies, do so in a fume hood. When opening the culture bottle after ether has been applied to it, position the mouth of the bottle away from you so you do not inhale the nasty fumes.) Place the maggot in the drop of saline.

4. Under the dissecting microscope, orient the narrow end of the larva to your dominant hand. Bear down on the mandible area (black hooks) with a probe or closed tip of fine forceps so that the maggot is held down firmly on the slide. With your free hand, use fine forceps to pull away with a quick motion the posterior one-third of the maggot until the body wall snaps just behind the mandibles (see Figure 15.1).

5. Note the paired, long, saclike salivary glands trailing posteriorly from the head end of the larva. They are translucent, slightly glossy, and flanked by a narrow sliver of fatty tissue on their outer sides. Remove all the debris in your saline puddle, leaving only the severed glands. Carefully dissect away and discard the fatty tissue flanking the salivary glands.

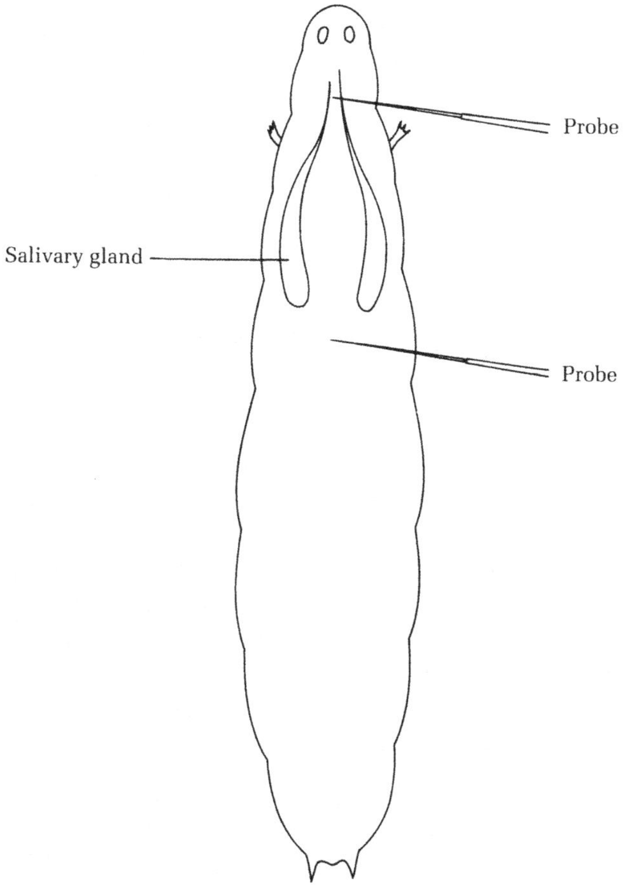

FIGURE 15.1

Dissecting a last-instar *Drosophila* larva to obtain salivary glands or imaginal discs. Hold larva head down with forceps or probe applied against a glass slide on which larva lies in a drop of insect saline. With a second probe or pair of forceps, hold larva at a point a third of the way down from its anterior end. Firmly pull the probes or forceps apart. Salivary glands should remain attached to the mouthparts, from which they should be carefully teased apart and freed of attached tracheoles and fat. Imaginal discs will also be visible: many will remain attached to viscera and integument by tracheoles; others will detach and fall to the surface of the glass slide.

6. While peering down into the dissecting microscope, use an absorbent paper square to draw off nearly all the fluid around the glands. You must be careful not to let them dry or to cause them to stick to the paper.

7. Immeditely add a drop of acetic acid to fix the glands. (What does "fix" mean in this context?) Draw off the acid in 30 sec, and replace it with aceto-orcein stain.

8. Allow the glands to sit in the stain for 10 min. Make sure the puddle of stain does not dry up. Add more stain when necessary.

9. Place a coverslip carefully over the puddle, being careful not to trap bubbles or lose the glands. To do this, lower one edge of the coverslip to the left of the puddle. Place the tip of the left index finger on the glass slide to steady the lowered edge. Prop up the upended remainder of the coverslip with a probe, fine forceps, or similar thin instrument, and gently lower the cover slip so that it contacts the glass slide from the left to the right. (This can be demonstrated if needed.) Gently flame (do not boil) the preparation.

10. Enclose the slide in the tissue wad, making sure that the coverslip is centered. Place the slide parallel to the edge of the table, plant a thumb firmly and squarely on the wad of tissue directly above the coverslip, stand up, and press down very firmly on the coverslip. Do *not* move your thumb forcefully from side to side because this could break up the chromosomes into small bits. However, with the right touch, moving the coverslip or running the forceps points over the coverslip could generate just enough pressure to spread, but not fracture, the chromosome arms.

11. Quickly scan the slide under the compound microscope to see if the chromosome spreads are good. A good chromosome spread is one in which the polytene chromosomes have been compressed into a loose or disentangled rosette of arms.

12. If the chromosome spreads are poor, make another slide.

13. If the chromosome spreads are good, seal your mount by applying a narrow band of nail polish around the edges of the coverslip. Reinforce the seal by re-applying two or three more coats of nail polish, allowing each one to air-dry well.

QUESTIONS

1. Polytene chromosomes are found in nondividing cells only (for example, in salivary gland cells and intestine cells). Why not in actively dividing cells?

2. Why is it necessary to "fix" a specimen for microscopic observation?

3. Propose a likely reason why regions along the polytene chromosome (bands) stain more darkly than other spots (interbands).

References

Biroc, S. L. (1986). *Developmental Biology: A Laboratory Manual with Readings,* pp. 154–169. New York: Macmillan Publishing Co.

Bridges, C. B. (1935). Salivary chromosome maps with a key to the banding of the chromosomes of *Drosophila melanogaster. J. Hered.* **26,** 60 and supplement.

Lefevre, G., Jr. (1976). A photographic representation and interpretation of the polytene chromosome of *Drosophila melanogaster* salivary gland. In *The Genetics and Biology of* Drosophila, Vol. 1a, pp. 31–66 (M. Ashburner and E. Novitski, eds.). London: Academic Press.

Lindsley, D. L., and Zimm, G. G. (1992). *The Genome of Drosophila melanogaster.* San Diego: Academic Press.

NOTES TO THE PREPARATOR

1. Wild-type *Drosophila melanogaster* cultures can be obtained from most biological supply houses, such as Ward's Natural Science Establishment, Inc. (5100 West Henrietta Road, P.O. Box 92912, Rochester, New York 14692-9012) and Carolina Biological Supply Company (2700 York Road, Burlington, North Carolina 27215). Set up the cultures early and expand the colony to have one culture bottle per two students. Use only last-instar larvae; these are easy to identify because they crawl out of the medium and up the interior surfaces of the culture jar.

2. Insect saline and aceto-orcein stain are prepared as follows:

   Insect Saline

   1.87 g NaCl

   0.875 g KCl

   0.078 g $CaCl_2$

   250 ml distilled water

   Aceto-orcein Stain

   6 g orcein powder dissolved in a hot mixture of

   150 ml glacial acetic acid

   60 ml lactic acid

   90 ml distilled water

   Filter twice. This stain will store indefinitely at 4°C.

3. A current genome map showing the identified bands and gene loci in *Drosophila melanogaster* chromosomes is available from *Science,* the journal of the American Association for the Advancement of Science (1333 H Street, N.W., Washington, D.C. 20005).

4. Another source of genome maps is Lindsley, D. L., and Zimm, G. G. (1992). *The Genome of Drosophila melanogaster.* Academic Press, Inc. (1250 Sixth Avenue, San Diego, California 92101). Eight original polytene maps of C. B. Bridges, P. M. Bridges, and G. Lefevre are bound into this reference book and are also available separately as an eight-map packet for student use. Toll free ordering available: 1-800-321-5068 or FAX 1-800-336-7377.

NOTES

NOTES

NOTES

Induction of Chromosome Puffing with Heat Shock

Many organisms are able to withstand brief pulses of high temperatures. In recent years, the molecular basis of this clearly adaptive ability has been investigated in diverse organisms. Notable among these is *Drosophila melanogaster*, again for the reasons explained earlier concerning this fly's suitability for many biological investigations.

The heat-shock response appears to be mediated by heat-shock proteins (HSPs). The synthesis of HSPs is triggered by exposure to heat. This means that HSP genes become transcriptionally active. In polytene chromosomes, signs of such activity are manifested by an increased number of puffs, or points of unraveling, or other signs associated with DNA activity. Because *Drosophila* polytene chromosomes are convenient to study, they will be used as a system for quantifying the extent of cellular response to heat. Students will make chromosome squashes from *Drosophila melanogaster* larvae exposed for a short period to 37°C heat. Although this temperature is normal for homoiotherms, it is generally too high for poikilotherms (*Drosophila* is a poikilotherm).

The procedure for this exercise is essentially that of the last lab class. However, this time students will be asked to dissect larvae from several culture jars that have been placed (or not!) in a 37°C incubator. As before, each student should count the number of puffs and loops encountered when closely examining ten entire chromosome arms. Because you will not know beforehand if the larvae that have been dissected were exposed to heat shock, you will (hopefully) have no biases. Record all observations on the blackboard. The control data will come from the last lab's exercise on nonexposed flies. During the next lab period, control and experimental data will be pooled and their implications discussed.

Note that the means of evaluating HSP production is very crude; that is, by counting chromosome puffs. There is, however, some basis for this apparent madness. As long ago as the late 1970s, cytogeneticists reported *heat-induced change in the puffing patterns* of salivary gland chromosomes in *D. melanogaster.* Recall that the puffing of a band in a polytene chromosome indicates intense transcription activity. Almost immediately on exposure to heat shock, preexisting polytene chromosome puffs regressed and *NINE* (eat your heart out!) new puffs were detected (Ritossa 1962, 1963, 1964). Heat, therefore, apparently causes the coordinated induction of scattered gene loci. These events, in turn, suggest that the stimulation of gene transcription is the primary response to heat shock.

Propose a mechanism by which heat and other stresses can induce the coordinated transcription of scattered genes in *Drosophila.* Be ready to present your proposal during the next lab meeting.

PROCEDURE

For this lab exercise, familiarize yourself with the *Drosophila* genome map available in the laboratory. Such preparation will allow you to identify two groups of bands, 87A and 87C, on the right arm of chromosome *3*. This chromosome arm is most easily identified based on the shape of its telomere (or free) end. If you think you have located these loci, present a convincing case to the instructor. Include a summary of the results in your report.

QUESTIONS

1. What functions do HSPs perform in *Drosophila?* Relate the production of HSPs to the human febrile response to viral or bacterial infection. Would HSPs have any adaptive significance in this case?

2. Polytene chromosomes have also been used as analytical tools for studying the deleterious effects of carcinogens and ionizing radiation. Give reasons why polytene, and not other types of chromosomes, are ideal for this purpose.

References

Ashburner, M., and Bonner, J. J. (1979). The induction of gene activity in *Drosophila* by heat shock. *Cell* **17**, 241–154.

Biroc, S.L. (1986). *Developmental Biology: A Laboratory Manual with Readings.* pp. 154–169. New York: Macmillan Publishing Co.

Lindsley, D. L., and Zimm, G. G. (1992). *The Genome of Drosophila melanogaster.* San Diego: Academic Press, Inc.

Ritossa, F. M. (1962). A new puffing pattern induced by heat shock and DNP in *Drosophila. Experientia* **18**, 571–573.

Ritossa, F. M. (1963). New puffs induced by temperature shock, DNP and salicylate in salivary chromosomes of *Drosophila melanogaster. Drosophila Information Service* **37**, 122–123.

Ritossa, F. M. (1964). Experimental activation of specific loci in polytene chromosomes of *Drosophila. Exp. Cell Res.* **35**, 601–607.

NOTES TO THE PREPARATOR

1. Wild-type *Drosophila melanogaster* cultures can be obtained from most biological supply houses, such as Ward's Natural Science Establishment, Inc. (5100 West Henrietta Road, P.O. Box 92912, Rochester, New York 14692-9012) and Carolina Biological Supply Company (2700 York Road, Burlington, North Carolina 27215).

2. Heat-shock larvae by incubating culture jars for 70 min and 40 min in a 37°C incubator.

3. Insect saline and aceto-orcein stain are prepared as follows:

   Insect Saline

   1.87 g NaCl

   0.875 g KCl

   0.078 g $CaCl_2$

   250 ml distilled water

   Aceto-orcein Stain

   6 g orcein powder dissolved in a hot mixture of

   150 ml glacial acetic acid

   60 ml lactic acid

   90 ml distilled water

   Filter twice. This stain will store indefinitely at 4°C.

4. A current genome map showing the identified bands and gene loci in *Drosophila melanogaster* chromosomes is available from *Science,* the journal of the American Association for the Advancement of Science (1333 H Street, N.W., Washington, D.C. 20005).

5. Another source of genome maps is Lindsley, D. L., and
    Zimm, G. G. (1992). *The Genome of Drosophila melano-*
    gaster. Academic Press, Inc. (1250 Sixth Avenue, San
    Diego, California 92101). Eight original polytene maps
    of C. B. Bridges, P. M. Bridges, and G. Lefevre are bound
    into this reference book and are also available separately
    as an eight-map packet for student use. Toll-free order-
    ing available: 1-800-321-5068 or FAX 1-800-336-7377.

NOTES

NOTES

Drosophila Imaginal Discs and Ecdysone

The transformation of a blind, limbless, wormlike, mandibulate *Drosophila melanogaster* maggot into a large-eyed, winged, legged fly that survives exclusively on liquefied food requires an enormous reorganization of body tissues. Such reorganization takes place during the pupal stage, a seemingly inactive period for the organism but actually a time of intense morphogenetic activity. Many of the processes occurring during the larva–pupa–adult transition are influenced by hormones. *Ecdysone* is a steroid hormone that brings about the onset of moulting in arthropods (insects, crustaceans, and spiders). However, the outcome of the moult (whether a later larval instar, pupa, or imago) is determined not by ecdysone, but by the level of another hormone, *juvenile hormone,* present. Thus, a high juvenile hormone to ecdysone (JH/E) value results in yet another larval instar; a low JH/E value transforms a larva into a pupa, or a pupa into the imago.

One of the target tissues for ecdysone is the *imaginal disc* (Figure 17.1). An imaginal disc is a patch of cells from which adult structures develop. The cells are initially undifferentiated, but quickly assume different identities during metamorphosis. It is here that ecdysone plays a major role. During pupation, the imaginal discs evert (telescope out) into adult structures such as legs, antennae, and wings. Eversion is triggered or induced by ecdysone.

In this exercise, students will dissect *Drosophila* last-instar larvae, retrieve imaginal discs (Figure 17.2), and attempt disc eversion by exposing them to ecdysone for 24 hr. At the end of the experiment, the class will pool data and determine which imaginal discs responded to any or all of three concentrations of ecdysone. The class will then examine the data to see if there is evidence for the threshold concept of hormone action on embryonic tissues. In the lab report, include a summary of class results and explain why they do or do not support the threshold concept.

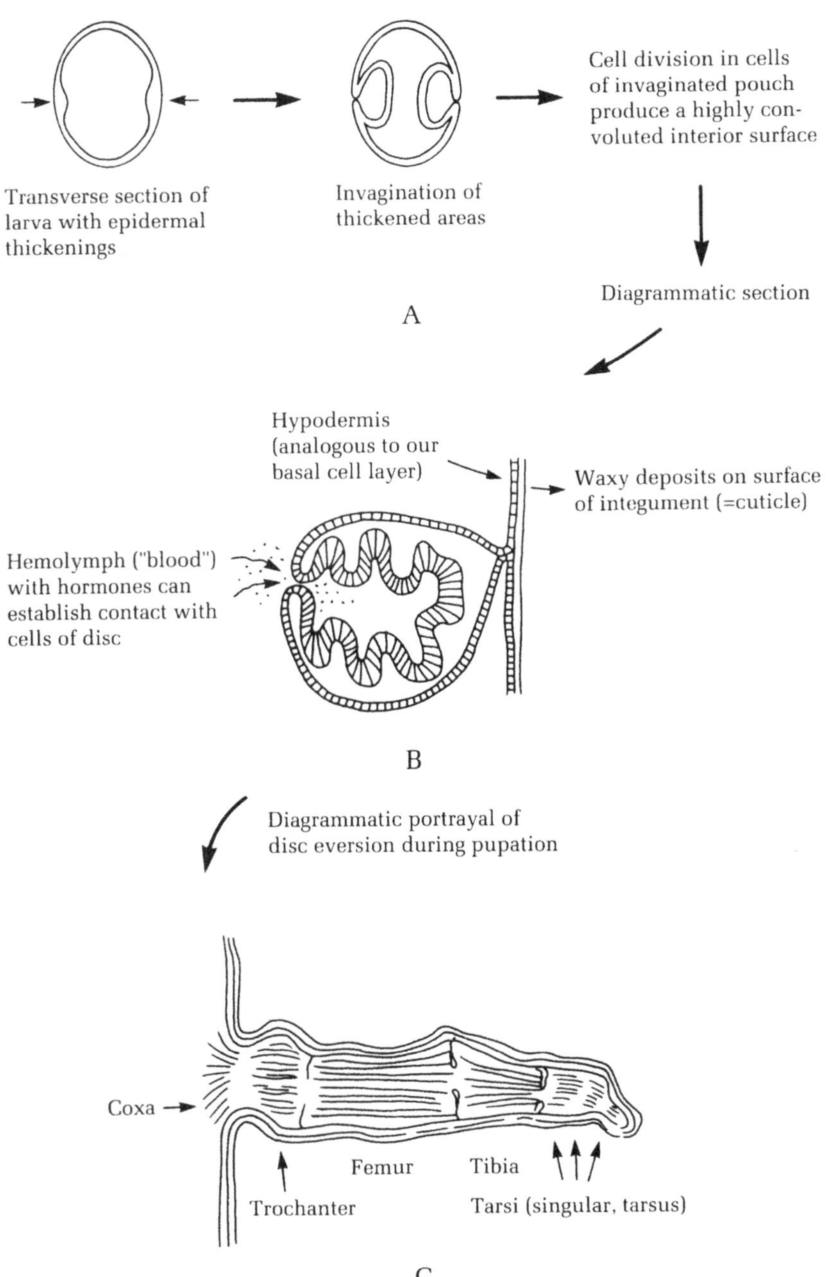

FIGURE 17.1

The imaginal disc originates as an invagination (A) of the larval body wall. Repeated division of cells in invaginated pouch produce highly convoluted interior surface (B). During the pupa–adult transition, imaginal disc everts (C). The musculature and innervation of the appendage, in this case a leg, eventually mature (D).

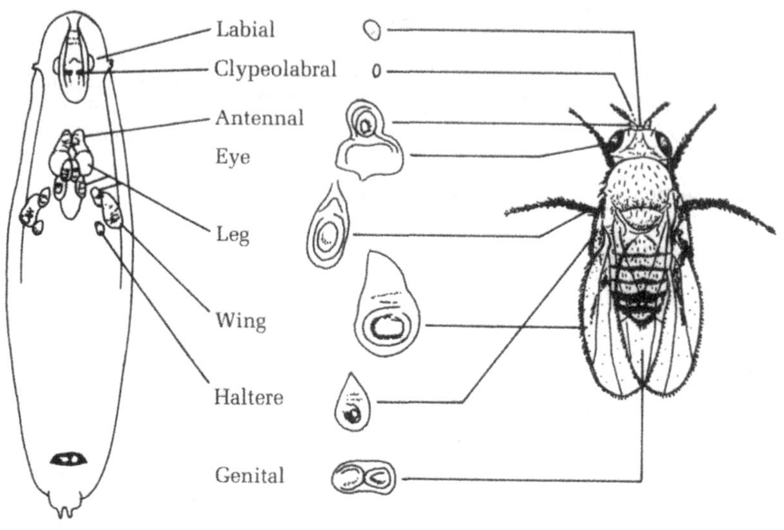

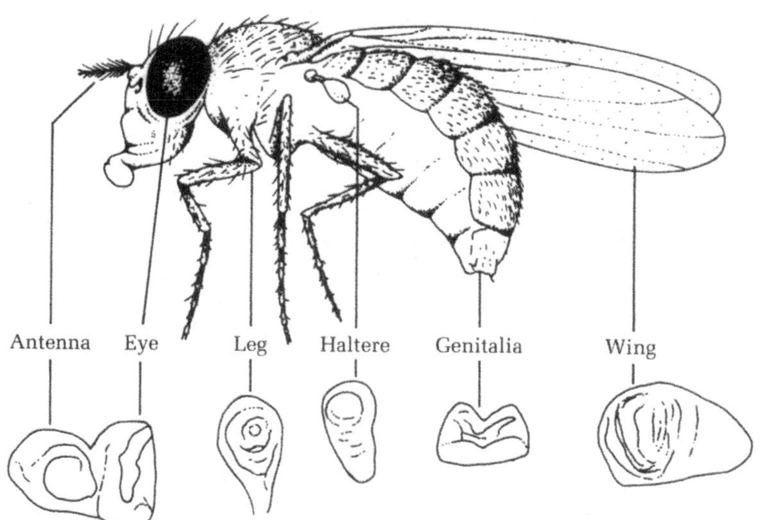

FIGURE 17.2

Locating and identifying imaginal discs. (A) Positions of different discs in the larva (left) and the body parts which they produce in the adult (right). (B) Lateral view of adult *Drosophila* showing body organs that form from commonly recovered, larger imaginal discs.

PROCEDURE

1. Practice using a finely pulled Pasteur pipette with a rubber bulb attached to aspirate small particles suspended in or lying under a drop of liquid on a glass slide. *Gradually* collapse or release the rubber bulb in order to get a weak vacuum; otherwise, the specimen could be lost.

2. Prepare 4 35-mm culture dishes by adding 2 ml of serum-supplemented Schneider's medium to each dish. (Use aseptic technique.) To one plate, add 5 µl of ecdysone solution (1 mg powder per ml 95% ethanol). To the second plate, add 2 µl of ecdysone solution; to the third, 1 µl. The fourth (control) plate should receive 5 µl of 95% ethanol.

3. Obtain a large (last-instar) maggot and rinse it clean of superficial debris in one or two drops of saline on a glass slide. Transfer the maggot to another drop of saline that has been formed on the floor of a sterile Petri dish. Sever the maggot using the same process as for obtaining salivary glands. Many of the imaginal discs are found in the anterior third of the animal. (See Figure 15.1 in Exercise 15.) The two wing and the six leg discs are the largest, and generally work best for this experiment. Some experimentalists claim that eye discs work well too, particularly if the strain of fly used has pigmented eyes, because pigment develops as the imaginal eye disc everts. Dissect the discs free of adhering tissue and allow them to sink to the bottom of the saline puddle.

4. Dissect enough maggots to give 20 to 25 discs.

5. With sterile, pulled Pasteur pipettes, aspirate 5 discs into each of the prepared Petri dishes (with Schneider's medium and ecdysone or alcohol). Examine each dish to make sure the discs were actually transferred. Sketch the appearance of these discs in the note pages provided. Using Figure 17.1, try to identify the discs that have been explanted.

6. Set the discs in an unused part of the lab and leave for 20–24 hr at room temperature. If necessary, change the Schneider's medium solution once during the incubation period. Remember to add ecdysone or alcohol as necessary.

7. Be sure to return for the 24-hr observation. *Sketch the discs.* Have they telescoped out? Describe the appearance of the discs with words and illustrations. Fine structural details do not preserve well even in fixed specimens. Only if there is absolutely no time to complete the sketch should the specimens be fixed at once.

8. Fix observed discs by removing most of the culture medium (slowly, with a Pasteur pipette) and replacing it with 2% glutaraldehyde. Seal the dishes with a strip of stretched Parafilm (ask to have this trick demonstrated if needed). Refrigerate your fixed specimens.

> *Note:* Glutaraldehyde is toxic. Clean up spills immediately. Wash thoroughly in case of dermal contact. Do not dispose of glutaraldehyde solutions in the sink. Instead, pour glutaraldehyde-containing solutions into labeled jar in the hood. Discard plastic containers into the waste box in the hood.

QUESTIONS

1. Which imaginal discs appeared to be most vulnerable to ecdysone effects? Defend your answer.

2. Transdetermination occurs when imaginal discs are serially transferred from a donor maggot to a series of recipients. What does this suggest about the mechanism of transdetermination? The basis of homeotic mutations?

References

Biroc, S. L. (1986). *Developmental Biology: A Laboratory Manual with Readings*, pp. 154–169. New York: Macmillan Publishing Co.

Fristrom, D., and Fristrom, J. W. (1975). The mechanisms of evagination of imaginal disks of *Drosophila melanogaster*. I. General considerations. *Dev. Biol.* **43**, 1–23.

Fristrom, J. W., Fristrom, D., Fekete, E., and Kuniyuki, A. H. (1977). The mechanism of evagination of imaginal disks of *Drosophila melanogaster*. *Amer. Zool.* **17**, 671–684.

Gilbert, S. F. (1991). *Developmental Biology,* 3rd ed., pp. 699–706. Sunderland, Massachusetts: Sinauer Associates.

NOTES TO THE PREPARATOR

1. See Exercise 15 for information regarding *Drosophila* cultures, making saline and orcein stain, and identifying chromosomes.

2. Schneider's medium is available from Gibco Laboratories (3175 Staley Road, Grand Island, New York 14072). To evoke disc eversion, this medium should be supplemented so that the final solution is 10% fetal calf serum. Ecdysone and fetal calf serum are available from Sigma Chemical Company (P.O. Box 14508, St. Louis, Missouri 63178-9916).

NOTES

NOTES

INDEX

A

Aceto-orcein, 208, 219
Acrosome reaction, 12
Adhesive organ, 51, 52, 53
Air space, 113
Albumen, 113
Alcian blue, 190–192, 195
Alizarin red, 190–192, 195
Allantois, 87, 98
Amnion, 98, 101, 113
Amniotic fold, 101
Animal hemisphere, 36
Animal pole, in embryo of
    frog, 38
    sea urchin, 25, 26
Anterior intestinal portal, 98, 102, 149
Anus
    in frog embryo, 36
    in sea urchin embryo, 24
Archenteron
    in frog, 36, 38, 40
    in sea urchin, 24, 26
Area opaca, 82
Area pellucida, 82

B

Bauchstucke, 66
Bipinnaria, 25, 27
Blastocoel, in embryo of
    chick, 82
    frog, 36, 38
    mouse, 177
    sea urchin, 24, 25, 26

Blastocyst, 177
Blastoderm, 82, 113, 140
Blastodisc, 82
Blastomere
    chick, 82
    frog, 36, 42, 44
    sea urchin, 24, 30
Blastopore
    dorsal lip of, 38
    of frog embryo, 36, 40, 66
Blastula
    frog, 36
    sea urchin, 17, 24
Brain
    in chick, 100
    in frog embryo, 51, 53, 54, 55

C

Caffeine, 124
Calcium ionophore, 12–13, 16, 19, 21
Candling box, 120
Cardia bifida, 148–156
Caudal flexure, 125
Cerebral hemisphere, 125
Chalaza, 113
Chick
    allantois, 87
    amnion, 98, 101, 113
    amniotic fold, 101
    anterior intestinal portal, 102, 149
    area opaca, 82
    area pellucida, 82
    blastocoel, 82
    blastoderm, 82, 113, 140

Chick (*continued*)
blastodisc, 82
blastomere, 82
brain, 100
cardia bifida, 148–156
caudal flexure, 125
cerebral hemisphere, 125
chorioallantoic membrane, 113
chorion, 98, 101
cleavage pattern, 82
coelom, 89
descending aorta, 102
dorsal aorta, 89, 101, 102
ear vesicle, 98, 100
embryonic coelom, 101
embryonic shield, 82
endocardium, 149
endoderm, 87
epiblast, 82–83
epidermis, 89
epimyocardium, 149
extraembryonic coelom, 101
eye, 98
eye vesicle, 88
forebrain, 89, 98, 102
foregut, 87, 102
forelimb bud, 100
ganglion, 98, 100, 102
head mesenchyme, 102
head mesoderm, 87
heart, 88, 89, 98
Hensen's node, 83
hindbrain, 89, 98
hindlimb bud, 100, 125
hypoblast, 82–83
integument, 89
intestinal portal, 88, 98
lateral body fold, 101
lens, 100
lens vesicle, 102
liver diverticula, 102
mandibular arch, 102
mesoderm, 89
mesonephros, 98, 100
midbrain, 99, 102
neural fold, 88, 89
neural plate, 88, 89, 98
nose rudiment, 100
notochord, 87, 88, 101, 102
olfactory pit, 125
optic cup, 100, 102, 125
otic vesicle, 125
pharynx, 89, 102
primitive groove, 83
primitive knot, 83
primitive streak, 82, 88, 98
pronephros, 101
sinus venosus, 102
somite, 88, 89, 98, 100, 102
spinal cord, 88, 89, 98, 100, 102
stomach, 98, 102
subgerminal cavity, 82
tail, 100
ventricle, 102
vitelline artery, 89, 125
vitelline vein, 88, 102, 125
vitelline vessel, 102
yolk sac, 87, 98, 101
Chordamesoderm, in frog embryo, 37
Chorioallantoic membrane, 113
Chorion, 98, 101
Chromocenter, 200
Cilia, 24
Cleavage
holoblastic, 82
in frog embryo, 36
in sea urchin embryo, 24
meroblastic, 82
Cloaca, 53
Coelom
in chick, 89
in sea urchin embryo, 25
Coelomic pouch, 25–26
Compound microscope, *see* Microscope
use
Cortical granules, 12

D

DMSO (dimethyl sulfoxide), 15
Descending aorta, 102
Dorsal aorta, 89, 101, 102

Dorsal fin fold, 51, 53, 56
Dorsal lip of blastopore, *see* Blastopore
Drosophila melanogaster
    chromocenter, 200
    chromosomes, 200–201, 216–217
    cultures, 208, 219
    ecdysone, 224–228
    genome map, 219–220
    heat-shock response, 216–217
    imaginal disc, 224–226
    juvenile hormone, 224
    salivary gland, 203
    somatic pairing, 200

E

Ear vesicle
    in chick, 98, 100
    in tadpole, 51, 52, 54
Ecdysone, 224–228
Echinoderm, 12, 24
Ectoderm
    in frog, 37
    in sea urchin, 24–25
Egg shell, 113
Embryogenesis
    in chick, 82–109
    in frog, 36–40
    in mouse, 177
    in sea urchin, 24–27
    in starfish, 24–27
Embryonic coelom, 101
Embryonic shield, 82
Endocardium, 149
Endoderm
    in chick, 87
    in frog, 37
    in sea urchin, 24–27
Epiblast, 82
Epiboly, 36
Epidermis
    in chick, 89
    in tadpole, 51
Epimyocardium, 149
Esophagus, 55
Ethyl alcohol, 124

External gills, 53, 54
Extraembryonic coelom, 101
Eye vesicle
    in chick, 88
    in tadpole, 51

F

Fertilization
    envelope, 12, 15, 17, 18
    external, 12
    in sea urchin, 12
Filopodia, 24
Forebrain
    in chick, 89, 98, 102
    in tadpole, 51
Foregut, 87, 102
Forelimb bud, 100
Frog
    adhesive organ, 51, 52, 53
    animal hemisphere, 36
    anus, 36
    archenteron, 36, 38, 40
    blastomere, 42, 44
    blastopore, 36, 40
    brain, 51, 53, 54, 55
    chordamesoderm, 37
    cloaca, 53
    development, time sequence for, 75
    dorsal fin fold, 51, 53, 56
    ear vesicle, 51, 52, 54
    ectoderm, 37
    endoderm, 37
    epiboly, 36
    epidermis, 51
    esophagus, 55
    external gills, 53, 54
    eye vesicle, 51
    forebrain, 51
    gall bladder, 55
    gastrulation, 36
    gill arches, 53
    gray crescent, 36, 42
    heart, 51, 54
    hindbrain, 52
    hindgut, 51, 56

Frog (*continued*)
    hypophysis, 51
    integument, 52, 53, 54, 55
    internal gills, 54
    intestine, 53, 55, 56
    involution, 36
    jelly coat, 36
    liver, 53, 55
    lung rudiment, 55
    mesoderm, 37, 44
    mesonephric duct, 56
    midbrain, 51
    midgut, 52
    mouth, 36, 51, 53
    myotome, 51, 52, 55
    neural crest cells, 52
    neural fold, 37, 40
    neural plate, 40
    neural tube, 37, 40
    nose rudiment, 51
    notochord, 37, 44, 51, 52, 53, 54, 55,
        56
    oral cavity, 51, 53
    pancreas, 55
    pericardial cavity, 52
    pharyngeal pouch, 51
    pharynx, 52
    pronephros, 51, 53, 55
    sperm entry point, 36
    spinal cord, 51, 52, 53, 55, 56
    stomach, 53, 55
    vegetal hemisphere, 36
    ventral fin fold, 51, 53, 56
    yolky endoderm, 51, 52

G

Gall bladder, 55
Gallus gallus, 112
Ganglion, 98, 100, 102
Gastrulation
    in frog, 36
    in sea urchin, 17
Gill arches, in tadpole, 53
Gray crescent, 36, 38, 66

H

Head mesoderm, 87
Heart
    in chick, 88, 89, 98
    in tadpole, 51, 54
Hensen's node, 83, 87
Hindbrain
    in chick, 89, 98
    in tadpole, 52
Hindgut, 51, 56
Hindlimb bud, 100, 125
Hypoblast, 82
Hypophysis, 51

I

Imaginal disc, 224–226
Integument
    in chick, 89
    in tadpole, 52, 53, 54, 55
Intestinal portal, 88
Internal gills, 54
Intestine, 53, 55, 56
Involution, 36–37

J

Jelly coat
    of frog embryo, 71
    of sea urchin embryo, 12
Juvenile hormone, 224

K

Krebs-Ringer solution, 184

L

Lateral body fold, 101
Lens, 100
Lens vesicle, 102

Liver
in chick, 102
in tadpole, 53, 55
Lung rudiment, 55
Lytechinus pictus, 21

M

Macromeres, 24, 25
Mandibular arch, 102
Mesenchyme
head, in chick, 102
primary, in sea urchin, 24, 26
secondary, in sea urchin, 24, 26
Mesoderm
in chick, 87, 89
in frog, 37, 44
in sea urchin, 26
Mesomeres, 24, 25
Mesonephros
in chick, 98, 100
in tadpole, 56
Micromeres, 24, 26
Microscope, compound, 4–5
eyepiece, 4
field diaphragm, 4
objective, 4
nosepiece, 5
ocular, 4
stage, 4
substage condenser, 4
substage diaphragm, 4
use, 3
zoom knob, 5
Midbrain
in chick, 98, 102
in tadpole, 51
Midgut, 52
Morula, 177
Mouse embryo, 177
Mouth
in sea urchin embryo, 24
in tadpole, 36, 51, 53
Myotome, 51, 52, 55

N

Neural crest cells, 39, 52
Neural fold
in chick, 88, 89
in frog embryo, 37, 40
Neural plate
in chick, 88, 89, 98
in frog embryo, 40
Neural tube, 37, 40
Neurulation, 39
Nose rudiment
in chick, 100
in tadpole, 51
Notochord
in frog embryo, 37, 44, 51, 52 53, 54,
55, 56
in chick, 88, 101, 102

O

Olfactory pit, 125
Optic cup, 100, 102, 125
Oral cavity, 51, 53
Organogenesis, 50
Otic vesicle, 125

P

Pancreas, 55
Parthenogenesis, 16
Pericardial cavity, 52
Pharyngeal pouch, 51
Pharynx
in chick, 89, 102
in tadpole, 52
Pluteus, 17, 27
Polyspermy, 12
Polytene chromosomes, 200–207
Primitive groove, 83
Primitive knot, 83
Primitive streak, 82, 88, 98
Pronephros
in chick, 101
in tadpole, 51, 53, 55

R

Rana pipiens, 46, 62
Rana catesbeiana, 162

S

Saline, insect, 208, 219
Salivary gland, 203
Scientific paper, 6–7
    abstract, 6
    conclusions, 7
    discussion, 7
    introduction, 6
    materials and methods, 7
    results, 7
    summary, 7
Sea urchin
    archenteron, 26
    animal pole, 25
    acrosome reaction, 12
    anus, 24
    blastocoel, 24, 25
    blastomeres, 24, 30
    blastula, 17, 24
    cilia, 24
    coelom, 25
    coelomic pouch, 25–26
    cortical granules, 12
    development, time sequence for, 17
    ectoderm, 24
    embryogenesis, 24–27
    endoderm, 26
    fertilization, 12
    fertilization envelope, 12, 17, 18
    gastrula, 17
    jelly coat, 12
    macromeres, 24, 25
    mesoderm, 26
    mesomeres, 24, 25
    micromeres, 24, 25
    mouth, 24
    plasmalemma, 12
    pluteus, 17, 27
    polyspermy, 12
    primary mesenchyme, 24, 26, 30
    secondary mesenchyme, 24, 26, 30
    spicules, 29
    vegetal plate, 24
    vegetal pole, 25
    zygote, 12
Shell membrane, 113
Sinus venosus, 102
Somatic pairing, 200
Somite, 88, 89, 98, 100, 102
Sperm entry point, 36
Spicules, 29
Spinal cord
    in chick, 89, 98, 100, 102
    in tadpole, 51, 52, 53, 55, 56
Starfish, 24
Stomach
    in chick, 98, 102
    in tadpole, 53, 55
Strongylocentrotus purpuratus, 21
Subgerminal cavity, 82

T

Tadpole, *see* Frog
Threshold concept, 162
Thyroxine, 162–166
Totipotency, 138, 176
Tyrode's solution, 133

U

Ultraviolet light, 66–69

V

Vegetal hemisphere, 36
Ventral fin fold, 51, 53, 56
Ventricle, 102
Vegetal plate, 24

Vegetal pole, in embryo of
    frog, 38
    sea urchin, 25
Vitelline
    artery, 89, 128
    envelope, 113
    vein, 88, 102, 128
    vessel, 102

X

Xenopus laevis, 46, 62, 70–72, 75, 76

Y

Yolk, 113
Yolk plug, 40
Yolk sac, 87, 98, 101
Yolky endoderm, 51, 52

Z

Zona pellucida, 177
Zygote, 12

Made in the USA
Monee, IL
07 July 2026

56545368R10142